Drug Metabolite Isolation and Determination

METHODOLOGICAL SURVEYS IN BIOCHEMISTRY AND ANALYSIS

Series Editor: Eric Reid

Guildford Academic Associates
72 The Chase
Guildford GU2 5UL, United Kingdom

The series is divided into Subseries A: Analysis, and B: Biochemistry
Enquiries concerning Volumes 1–11 should be sent to the above address.

Drug Metabolite Isolation and Determination

Edited by

Eric Reid

and

J. P. Leppard
Formerly of the University of Surrey

PLENUM PRESS • NEW YORK AND LONDON

Library of Congress Cataloging in Publication Data

Main entry under title:

Drug metabolite isolation and determination.

(Methodological surveys in biochemistry and analysis. Subseries A, Analysis; v. 12)
Based on the Fourth International Bioanalytical Forum, held Sept. 7–10, 1981, at the University of Surrey, Guildford, United Kingdom.
Includes bibliographical references and indexes.
1. Drugs—Analysis—Congresses. 2. Drugs—Metabolism—Congresses. I. Reid, Eric, 1922–   . II. Leppard, J. P. III. International Bioanalytical Forum (4th: 1981: University of Surrey) IV. Series.
RS189.D784 1983   615.7′028   83-2214
ISBN 978-1-4684-4486-5   ISBN 978-1-4684-4484-1 (eBook)
DOI 10.1007/978-1-4684-4484-1

Based on the Fourth International Bioanalytical Forum held September 7–10, 1981, at the University of Surrey, Guildford, United Kingdom

© 1983 Plenum Press, New York
Softcover reprint of the hardcover 1st edition 1983

A Division of Plenum Publishing Corporation
233 Spring Street, New York, N.Y. 10013

Senior Editor's Preface

In drug metabolism, as in kindred fields, sound conclusions hinge on good methodology with special skills which, however, have had less than their share of recognition and 'kudos'. This series of books is earning regard for its description and encouragement of good methods, not least because of emphasis on difficulties and on rationale rather than recipes such as prevail in the literature.

Each book is based on a Forum — in this case the 4th Bioanalytical Forum held at the University of Surrey in September 1981 — and mirrors some of the cut-and-thrust of the debates, but is not to be regarded as a symposium record. There has in fact been strong editing, partly for the sake of clarity and of relevance to analytical practice (thus, pharmacokinetic data have been excluded). The aim has been to produce a balanced desk book, well cross-referenced and complementing earlier 'Analysis' volumes, whose material on particular analytes is cumulatively re-indexed here (Index of Compound Types); the concluding article and the General Index aid retrieval of lore on problems and approaches.

The evolution of the *Methodological Surveys* twin series (outline facing title page) has undergone yet another discontinuity, in respect of publisher and format. For Vols. 10 and 11, conventional typesetting by the publisher was obligatory; it led to a generally good appearance, although there were errors that the page-proofing should have eliminated, but it contributed to the slowness of publication. Moreover, it entailed an intolerable editorial burden, now reduced by reversion to 'camera-ready' text produced at the Guildford end. The initiative of switching to Plenum Press, with marketing and other strengths and with pricing that appeals to American and other customers, has entailed sad parting from the U.K. publisher whose involvement could have started with Vol. 5 (illness precluded this) and did start with Vol. 6. If the now widening readership can get past volumes back-ordered, this will help not only bibliographic searching but also future Forum financing. Here too there is a discontinuity: from 1983 each Forum will be under the auspices of a new Trust (an 'educational charity'), although the normal venue will still be the nearby campus of the well-liked University of Surrey. Hopefully there will still be company support, which for the 1981 Forum came from Ciba-Geigy (U.K.), Glaxo and ICI Pharmaceuticals. It has been a boon that many contributors made little claim on Forum funds.

Scope of the book, and acknowledgements.- The articles are focused on 'real problems' in body-fluid analysis, typically with a final chromatographic separation of µg or often ng amounts if the aim is quantitation.   The pitfalls may not be realized by a typical chemist (cf. remarks in #A-3), but he may excel in metabolite identification – which this book covers to a fair extent. Where identity is known, and the metabolite could interfere in therapeutic drug monitoring or in diagnosis or itself have clinical relevance, useful guidance will come from articles that follow.   Authors have gone to much trouble, and are not to be blamed by any reader who would have liked an introduction to chromatography or to metabolic pathways (cf. list of conjugation reactions at end of concluding article). Appreciation is also expressed for permission to reproduce published material; the acknowledged sources include *J. Chromatog.* (Elsevier; e.g. in #A-1), *Anal. Chem.* (American Chemical Society; #A-2) and Wiley.

HPLC nomenclature.- With regard to the terms 'NP' and 'RP' (normal-, reverse-phase) the Editors have respected authors' preferences (cf. comment following Table 2 in #A-2) but share misgivings expressed by J.H. Knox and (in *LC in Practice)* by P.A. Bristow, who writes: "Reverse phase is a term which could well be retired gracefully".   Thus, an 'uncapped' packing with a low content of bonded alkyl groups may behave adsorptively.   Some authors exclude unbonded ('straight') silica from the 'NP' category that includes polar liquid stationary phases and polar bonded phases such as cyano. The designations 'hydrophobic' and 'hydrophilic' chromatography (Knox) have merit, and the concept of 'surface zone' rather than 'phase'. Some comment on so-called ion-pairing is made in #E (cf. p. 91).

Various abbreviations.- Nowadays the terms TLC (HPTLC = 'high performance' variant) and HPLC need no definition. In GC (gas-liquid in the present context), detector types include FID = flame ionization, AFID = alkali flame ionization ('nitrogen detector') which can detect phosphorus (hence NPD, N-PD), and ECD = electron capture but unfortunately connoting electrochemical detection in HPLC work. Mass spectrometry is denoted MS; EI = electron-impact, and CI = chemical ionization.   In HPLC, t_R (preferred; otherwise R_t) = retention time.

Non-chromatographic abbreviations include i.s. = internal standard (usually 'processed'; see #A-6 in Vol. 7), RIA = radioimmunoassay, UV = ultraviolet [absorptiometry], IR = infrared, and NMR = nuclear magnetic resonance.

An example of metabolite jargon is a dialkylated amine drug, 'D', that has lost one alkyl group: it might variously be termed de(s)alkylD, monodesalkylD, desmonoalkylD, norD...... A conjugate is 'Phase II'.

Guildford Academic Associates   ERIC REID
72 The Chase, Guildford GU2 5UL
Surrey, U.K.   1 April 1982

Contents

The NOTES & COMMENTS ('NC' items) at the end of each section comprise
some comments made at the Forum on which the book is based, together
with some supplementary points besides those in the concluding survey.

[†]*Sub-listed here: comments on particular articles in the section;
likewise on pp. 135, 189 & 231.*

---
* *Analytical case histories*

List of Authors

'Notes' (headed 'NC......' in the text) are not distinguished from main contributions.

C.D. Bevan – pp. 111–118
as for J. Chamberlain

A. Bhatti – pp. 181–187
as for J. Chamberlain

G.R. Bourne – pp. 97–102
ICI Pharmaceuticals, U.K.

M.A. Brooks – pp. 201–206, 225–230
as for J.A.F. de Silva

T.A. Bryce – pp. 181–187
as for J. Chamberlain

J. Caldwell – pp. 161–179
St. Mary's Hosp. Med. School,
London

J. Chamberlain – pp. 111–118,
  181–187, 249
Hoechst Pharmaceutical Res. Labs.,
U.K.

Anne Chambers – pp. 103–110
as for M.J. Stewart

F.G. Coppin – p. 249
as for J. Chamberlain

H. de Bree – p. 69
Duphar Res. Labs., The Netherlands

D. Dell – p. 249
Hofmann–La Roche, Switzerland

J.A.F. de Silva – pp. 201–206,
  225–230.
Hofmann–La Roche, U.S.A.

R.A. de Zeeuw – pp. 75–80
State Univ., Groningen,
The Netherlands

W. Dieterle – pp. 13–22
Ciba–Geigy, Switzerland

B.F.H. Drenth – pp. 75–80
as for R.A. de Zeeuw

C.V. Eadsforth – pp. 119–133
Shell Res., U.K.

J.W. Faigle – pp. 13–22
as for W. Dieterle

Y.Y.Z. Farid – pp. 141–142
as for M.J. Stewart

R.T. Ghijsen – pp. 75–80
as for R.A. de Zeeuw

B.L. Goodwin – pp. 47–53, 71–73
Queen Charlotte's Hosp., London

M.R. Hackman – pp. 201–206
as for J.A.F. de Silva

N.G.L. Harding – pp. 141–142
as for M.J. Stewart

M.P. Harrison – pp. 191–196
ICI Pharmaceuticals, U.K.

A.J. Hutt – pp. 161–179
as for J. Caldwell

J.B. Houston – pp. 207–214
Univ. of Manchester, U.K.

H.P.A. Illing – pp. 181–187
as for J. Chamberlain

T. Jagersma – pp. 75–80
as for R.A. de Zeeuw

D.R. Jarvie – pp. 103–110
Royal Infirmary, Edinburgh, U.K.

C.R. Jones - pp. 33-40
Wellcome Res. Labs., U.K.

L.J. King - pp. 247-248
Univ. of Surrey, U.K.

Mary V. Marsh - pp. 161-179
as for J. Caldwell

S.R. Moss - pp. 97-102
as for G.R. Bourne

H. Munzer - pp. 215-223
as for J. Rosenthaler

R.G. Muusze - pp. 241-242
Algemeen Psychiatrisch Inst.,
Delft, The Netherlands

E.A.M. Neill - pp. 243-246
Wellcome Res. Labs., U.K.

F. Overzet - pp. 75-80
as for R.A. de Zeeuw

J. Pao - pp. 225-230
as for J.A.F. de Silva

E. Reid - pp. 255-270
Univ. of Surrey, U.K.

J.C. Rhodes - pp. 207-214
as for J.B. Houston

W. Ritter - pp. 55-62, 231-238
Bayer, Wuppertal, W. Germany

D.R. Roberts - pp. 81-88
ICI Pharmaceuticals, U.K.

J.D. Robinson - pp. 111-118
as for J. Chamberlain

J. Rosenthaler - pp. 215-223
Sandoz, Basle, Switzerland

K.A. Sinclair - pp. 161-179
as for J. Caldwell

G.G. Skellern - pp. 3-11
Univ. of Strathclyde, U.K.

M.J. Stewart
Royal Infirmary, Glasgow, U.K.

D.A. Stopher - pp. 65-67
Pfizer Central Res., U.K.

Jelka Tomašić - pp. 137-139, 149-160
'Ruder Bošković' Inst., Yugoslavia

R. Voges - pp. 215-223
as for J. Rosenthaler

I.D. Watson - pp. 103-110, 141-142
Royal Infirmary, Glasgow, U.K.

R.E. Weinfeld - pp. 201-206
as for J.A.F. de Silva

I.D. Wilson - pp. 111-118, 181-187
as for J. Chamberlain

Section #A

TECHNIQUES APPLICABLE TO METABOLITE INVESTIGATION

#A-1

STRATEGIES CENTERED ON HPLC

G.G. Skellern

Drug Metabolism Research Unit
Department of Pharmaceutical Chemistry
University of Strathclyde
Glasgow G1 1XW, U.K.

The choice of an HPLC system will depend on its application and on the nature of the metabolites to be measured. Normal-phase (NP) chromatography is limited to the measurement of drugs whose physico-chemical properties are similar to the parent drug, whereas reverse-phase (RP) chromatography in its various forms has made possible the simultaneous measurement of metabolites varying widely in pK_a, lipophilicity and polarity. Either alone or in combination with other chromatographic techniques, HPLC has aided the isolation of some polar metabolites which may be difficult to isolate from biological material. In contrast with other types of chromatography, the in vitro formation of metabolites can be monitored directly by HPLC, with minimal sample preparation. Thus HPLC will aid the study of enzyme kinetics and reaction mechanisms.

The biotransformation of a compound may result in a diversity of metabolites with widely differing physico-chemical properties. It is therefore desirable to know the properties of the parent compound and its possible metabolites in order that a suitable HPLC system may be chosen. Its choice also depends on whether it is being used primarily as an analytical method for the identification of a metabolite and the determination of its concentration, or preparatively, when substantial quantities of one or more of the metabolites are to be isolated for further structural characterization. HPLC has been used extensively for studying the metabolism of drugs, particularly drugs of high relative molecular mass and polar metabolites [1].

Recently the 'State of the Art' of HPLC has been reviewed [2,

3], and the theoretical and applied aspects discussed, in addition
to newer developments.   Consequently I will focus upon particular
ways in which HPLC has been exploited in metabolic studies.

The subject of sample preparation and clean-up is discussed
by other contributors (e.g. # A-2, #E). It suffices to say that RP
materials are useful for sample clean-up and enrichment.   Thus, a
liquid-solid extraction with ODS material in cartridges removed the
sparingly water-soluble mebendazole and its metabolites from plasma
[4], yielding a chromatographically cleaner extract in a single
operation without substantial loss of compound. The authors state
that they could re-use the cartridges without loss of performance.

The first decision to be made is the type of chromatography to
be adopted.   NP (adsorption and bonded-phases) and RP systems are
applicable to the study of Phase I metabolites and methylated and
acetylated metabolites, whereas RP systems with either ion-pair or
ion-suppression modes are suitable for separating the polar, water-
soluble conjugates (glucuronides, sulphoconjugates and amino-acid
conjugates).   This article will concentrate more on the latter type.

SELECTION OF HPLC SYSTEM

Paracetamol is an example of a drug which is eliminated from
the body predominantly as its highly polar and water-soluble glucuronide and sulphate conjugates. The cysteine and *N*-acetylcysteine
conjugates are two other polar metabolites which are excreted in
significant amounts when a substantial overdose of drug (20 g) is
ingested. In this case filtered aliquots of urine can be directly

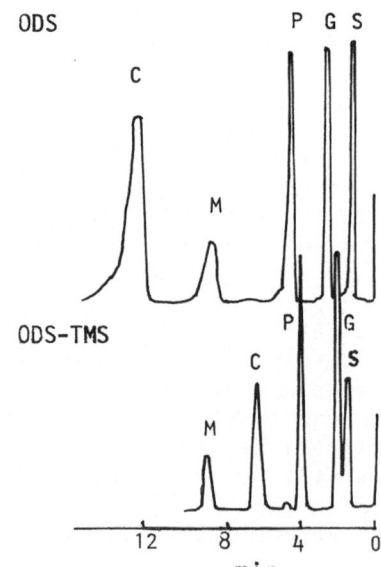

Fig. 1. Comparison of ODS silica
(upper) with ODS-TMS silica packing *(lower;* TMS = trimethylsilyl*)*
for separating paracetamol and
its metabolites ([5], *by permission).*
Eluant: water – methanol – formic
acid (86:14:0.1 v/v/v).
S = sulphate conjugate; G =glu-
curonide; P = paracetamol; C = cysteine conjugate; M = mercapturate.

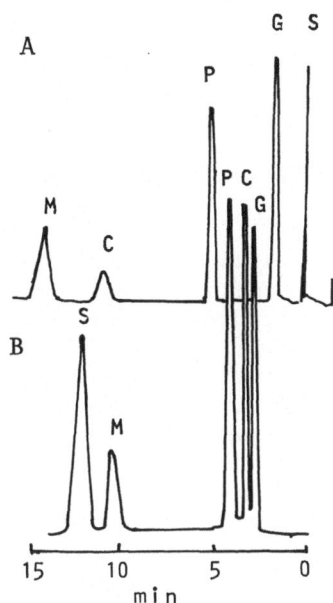

Fig. 2. Separation of paracetamol metabolites on ODS-TMS silica in the absence (**A**) and presence (**B**) of an ion-pairing agent ([6], *by permission*).
Eluant: **A** as for Fig. 1
B as for Fig. 1 but with dioctylamine (DOA; 0.7 mg /l) and KNO_3 (3 g/l).

injected onto the HPLC column, since there is a sufficient concentration of metabolites. With methanol-water-formic acid, Knox & Jurand [5] compared conventional ODS-silica with ODS-silica capped with TMS, and found the latter superior for the determination of paracetamol and its polar metabolites (Fig. 1). The effect of varying the concentration of methanol, formic acid and added salts on the k' values of the 5 compounds was thoroughly examined. The presence of acid in the eluting solvent markedly increased k' for these compounds, partly by suppressing the degree of ionization of the acidic metabolites. Acid in the eluant also decreased the possibility of hydrolysis of the conjugates. Slight pH changes of the eluant around the pK_a value of the metabolites can markedly alter k'. With straight methanol, water and formic acid mixtures as eluant, the elution order was sulphoconjugate, glucuronide, paracetamol, cysteine and *N*-acetylcysteine conjugates. However, upon the addition of salt (KH_2PO_4) to the eluant the capacity factor of the sulphoconjugate increased from zero to 0.8, and the elution order was glucuronide, sulphoconjugate, cysteine conjugate, paracetamol and mercapturate.

The addition of either dioctylamine (DOA) or tetrabutylammonium hydroxide (TBAOH) to the eluant to form hydrophobic ion-pairs with the highly ionized sulphate conjugate dramatically increased its k' value [6], the elution order now being glucuronide, cysteine conjugate, paracetamol, mercapturate and sulphoconjugate. To obtain reasonable values of k' for the mercapturate and sulphoconjugate it was necessary (Fig. 2) to add a salt (KNO_3) to the eluant when DOA

was used. However, there was not such a pronounced effect on k' for
the sulphoconjugate when TBAOH was added to the eluant, thus making
the addition of salt unnecessary. The column loadings for DOA and
TBAOH were up to 7 mg/l and 200 mg/l respectively.

     After optimization for the separation of the reference standards
the composition of the eluant may have to be modified when biological
material is being examined.

     The ability to use aqueous eluants has facilitated the isolation
and characterization of polar water-soluble metabolites. When ion-
suppression is used, then the eluate corresponding to the metabolite
peaks can be collected and either freeze-dried [7] or the solvent
removed under reduced pressure. If necessary the conjugate can then
be reconstituted and further purified by HPLC. The use of ion-pair
reagents in the eluant does not necessarily negate the use of MS,
although there may be a considerable excess of reagent relative to
the metabolite.   Thus, DOA because of its low eluant concentration
was preferred to TBAOH when MS was used to characterize the metabolites of paracetamol [6].

ISOLATION AND IDENTIFICATION OF METABOLITES

     Usually an enrichment step is required prior to the use of HPLC
for purification and isolation purposes.   For sample enrichment,
body fluids can be injected directly [8] onto fully automated HPLC
possessing pre-columns, which is useful for the routine determination
of drugs and metabolites when their structure is known.   However,
for identification purposes it may be necessary to extract the acidic
conjugate from the biological fluid prior to using HPLC. Probenecid
acyl glucuronide was identified by ^{13}C-NMR after its extraction from
acidified urine and purified with other metabolites of probenecid
using ODS material [9].   Similarly Veenendaal & Meffin[10] used this
approach for the purification and identification of the glucuronide
of clofibric acid.   The evidence that the purified compound was an
O-glucuronide was that it was hydrolyzed in the presence of β-glucuronidase to clofibric acid, and that this reaction was inhibited by D-
saccharo-1,4-lactone, a β-glucuronidase inhibitor. These reactions
were monitored by HPLC.   Further evidence of the identity of the
glucuronide was that its absorbance ratio, measured at 227 nm and
277 nm, was similar to the ratio observed for the aglycone, making
the reasonable assumption in this case that the glucuronyl moiety
did not alter the absorption spectrum. Plasma concentrations as low
as 1.5 µg/ml of glucuronide were directly measured by HPLC, by
adding phenolphthalein glucuronide as an internal standard, in a tri-
chloroacetic acid solution (Fig. 3).   Interestingly, 4 isomers of
the glucuronide of clofibric acid have been reported on the basis of
GC-MS results, viz. the α and β anomers of the pyranose and furanose
forms of the glucuronide [11].

The diastereoisomeric glucuronides of the 1,4-benzodiazepine, oxazepam, have been quantitatively measured by HPLC using ODS-silica and ion suppression [12], after their isolation and separation with XAD-2 and DEAE-cellulose-DE-23. The rate of enzymic hydrolysis of the purified S-(+) and R-(-) isomers was studied by HPLC with various β-glucuronidases from different sources. This study elegantly illustrated that this enzyme assay does not always provide unequivocal evidence of the identity of a suspected glucuronide if the production of aglycone is being measured. The rate of hydrolysis of the S-isomer was 400 times faster than for the R-isomer, for *E. Coli* β-glucuronidase, and moreover this enzyme preparation was far more stereoselective than β-glucuronidases obtained from bovine liver, marine molluscs and *Helix pomatia*.

Similar observations were made in another study [13] where HPLC was used to monitor the hydrolysis of 1-naphthol glucuronide and 1-naphthol sulphate by β-glucuronidases and aryl sulphatases from different sources.

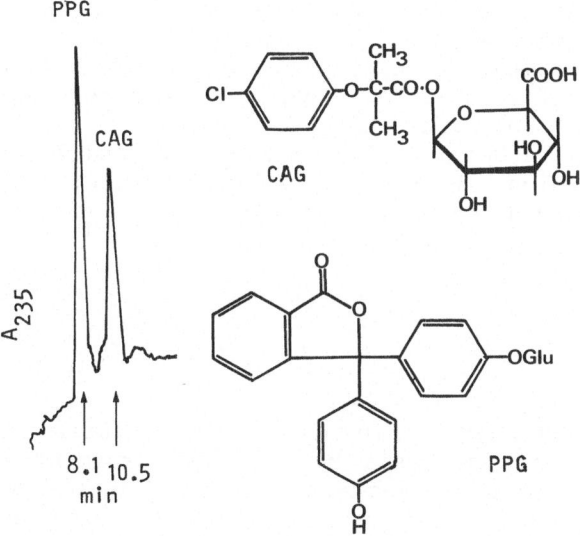

Fig. 3. Separation of the glucuronides of clofibric acid and phenolphthalein on ODS-silica [10].
Eluant: acetonitrile-glacial acetic acid-water (450:5:545).
CAG = clofibric acid glucuronide; **PPG** = phenolphthalein glucuronide.

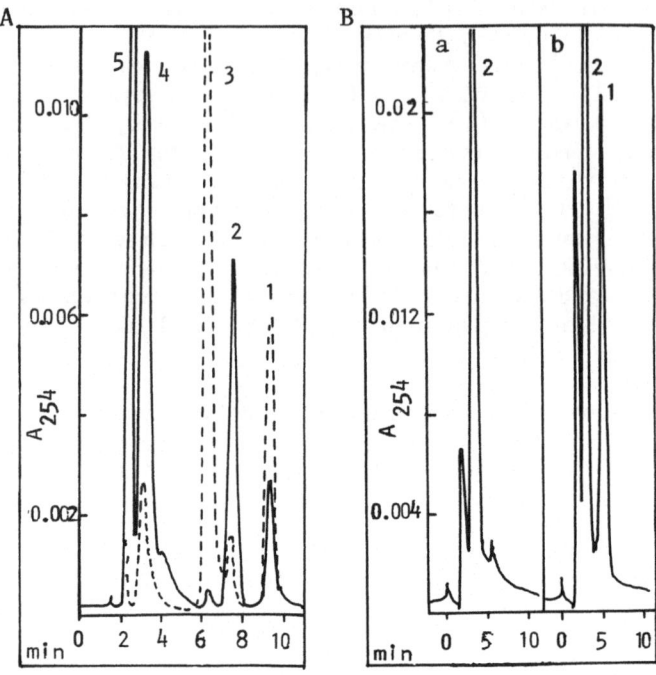

Fig. 4. Chromatograms recorded during *in vitro* glucuronidations
(from [18], *by permission*).
A *(left)*: paracetamol, at 2 min *(broken line)* and 30 min *(solid line)*
from the start of incubation. **Peaks: 1** = paracetamol; **2** = paracetamol
glucuronide; **3** = UDPGA; **4** = UMP; **5** = uridine.
B *(right)*: *p*-nitrophenol (PNP), at 2 min (**a**) and 20 min (**b**) from the
start of incubation. **Peaks: 1** = PNP glucuronide; **2** = UDPGA. Unconjugated PNP is retained on the column.

        Routinely HPLC has become the final purification stage for a meta-
bolite prior to its analysis by MS, mass fragmentography and NMR.
Using these spectroscopic methods a definitive statement can be made
about the structure of a metabolite. However, both on-line HPLC–NMR
[14] and on-line HPLC–MS [cf. #C-6 in Vol. 7, this series – *Ed.*] are
still in their infancy, the former in its gestation period! The more
widely used methods of HPLC detection (UV, fluorescence, electrochemical, radiochemical) provide limited valuable information. Recently
on-line HPLC–MS using RP chromatography and post-column ion-pairing

with a continuous-extraction interface has been evaluated for a mixture containing procainamide and *N*-acetylprocainamide, its metabolite, and *N*-propionylprocainamide and lignocaine [15]. In the presence of alkyl sulphonate counter-ions both electron and chemical ionization spectra were obtained.

Post-column reactors [cf. #D-5 in Vol. 7, this series - *Ed.*] may improve sensitivity and specificity as has been demonstrated for the detection of warfarin and its metabolites with a chemical reactor [16] and for purines with an enzymic reactor [17].

IN VITRO STUDIES

It was a reasonable extension for HPLC to be applied to the *in vitro* study of Phase I and Phase II biotransformations, since it is possible to inject biological fluids containing metabolites directly into the column, concentration permitting. Once a particular metabolite has been characterized, its rate and mechanism of formation can be studied with use of enzyme preparations. HPLC is ideal for this purpose, since it will not only measure either the substrate disappearance or the product's formation, but can monitor both simultaneously. The glucuronidation of paracetamol and *p*-nitrophenol by liver UDP-glucuronyltransferase has been measured by HPLC [18] by injecting aliquots of the incubation mixture into an ODS-silica column (Fig. 4, A & B). Buckpitt *et al.* [7] quantitatively determined and identified with HPLC the glutathione, cysteine and *N*-acetylcysteine conjugates of ^{14}C-paracetamol after it had been incubated with liver microsomal protein and an NADPH-generating system in the presence of either glutathione or cysteine or *N*-acetylcysteine. Metabolite fractions were also isolated and purified for chemical and MS analysis.

DIAGNOSTIC ROLE

That RP-HPLC has a predictive role with respect to the identification of conjugates is implicit in the aforementioned studies. Those familiar with GC will be aware of Kovats Indices and their predictive function. Based on the relative retention on ODS material of a homologous series of C_3-C_{23} 2-alkylketones, Baker [19] has constructed a retention index scale. The index of a drug/metabolite is calculated from the following equation:

$$\text{Retention Index} = 100 \; \frac{\log k'_x - \log k'_N}{\log k'_{N+1} - \log k'_N} \; + 100 \; N$$

where k'_x = capacity factor for drug/metabolite, k'_N = capacity factor for ketone standard eluting prior to drug/metabolite, k_{N+1} = capacity factor for the next higher homologue, N = no. of carbon atoms in the ketone.

Table 1. The Retention Index shift for glucuronides relative to their aglycones [20].

	Retention Index shift
Morphine 3-glucuronide	-264
Codeine glucuronide	-223
Testosterone glucuronide	-210
6-Bromonaphthol glucuronide	-254
Phenolphthalein glucuronide	-232
8-Hydroxyquinoline glucuronide	-301
p-Nitrophenol glucuronide	-223
Mean	-244 ±31

Using this equation the Retention Indices have been determined for 7 glucuronides and their aglycones using ODS-silica eluted with a phosphate buffer (pH 7)-methanol mixture [20]. An average negative shift in the Retention Index of 244 ±31 units was observed for the glucuronides (Table 1). Thus the use of Retention Indices might aid the identification of uncharacterized glucuronides and other metabolites. Their absorbance ratios were also determined by measuring their absorbance at 254 nm and 280 nm. Absorbance ratios may have predictive value since in some cases the introduction of a glucuronyl moiety into the molecule does not markedly change its 254 nm and 280 nm absorbance; in other glucuronides there was a marked difference in the absorbance ratio. The fact that the k' value for sulphate conjugates can be markedly changed upon the addition to the eluant of a basic ion-pairing reagent may be of diagnostic value.

HPLC has demonstrated its applicability and versatility in *in vivo* and *in vitro* metabolism studies, in particular in facilitating the isolation and characterization of highly polar acidic conjugates.

References

1. Skellern, G.G. (1981) *Analyst 106*, 1071-1075.
2. Several authors (1980) *J. Chromatog. Sci. 18*, 393-486.
3. Several authors (1980) *J. Chromatog. Sci. 18*, 487-582.
4. Allan, R.J., Goodman, H.T. & Watson, T.R. (1980) *J. Chromatog. 183*, 311-319.
5. Knox, J.H. & Jurand, J. (1977) *J. Chromatog. 142*, 651-670.
6. Knox, J.H. & Jurand, J. (1978) *J. Chromatog. 149*, 297-312.
7. Buckpitt, A.R., Rollins, D.E., Nelson, S.D., Franklin, R.B. & Mitchell, J.R. (1977) *Anal. Biochem. 83*, 168-177.

8. Roth, W., Beschke, K., Jauch, R., Zimmer, A. & Koss, F.W. (1981) *J. Chromatog. 222*, 13–22.
9. Eggers, N.J. & Doust, K. (1981) *J. Pharm. Pharmacol. 33*, 123–124.
10. Veenendael, J.R. & Meffin, P.J. (1981) *J. Chromatog. 223*,147–154.
11. Hignite, C.E., Tschanz, C., Lemons, S., Wiese, H., Azarnoff, D.L. & Huffman, D.H. (1981) *Life Sci. 28*, 2077–2081.
12. Ruelius, H.W., Tio, C.O., Knowles, J.A., McHugh, S.L., Schillings, R.T. & Sisenwine, S.F. (1979) *Drug Metab. Disp. 7*, 40–43.
13. Rhodes, J.C. & Houston, J.B. (1981) *Xenobiotica 11*, 63–70.
14. Bayer, E., Albert, K., Nieder, M., Grom, E. & Keller, T. (1979) *J. Chromatog. 186*, 497–507.
15. Kirby, D.P., Vouros, P., Karger, B.L., Hidy, B. & Petersen, B. (1981) *J. Chromatog. 203*, 139–154.
16. Lee, S.H., Field, L.R., Howald, W.N. & Trager, W.F. (1981) *Anal. Chem. 53*, 467–471.
17. Tawa, R., Kito, M. & Hirose, S. (1981) *Chem. Lett. 6*, 745–748.
18. Knight, B.I. & Skellern, G.G. (1980) *J. Chromatog. 192*, 247–249.
19. Baker, J.K. & Ma, C.Y. (1979) *J. Chromatog. 169*, 107–115.
20. Baker, J.K. (1981) *J. Liq. Chromatog. 4*, 271–278.

#A-2

PREPARATIVE LIQUID-CHROMATOGRAPHIC
METHODS IN DRUG METABOLISM STUDIES

W. Dieterle and J.W. Faigle

Research Department, Pharmaceuticals Division
CIBA-GEIGY Limited, CH-4002 Basle, Switzerland

The main difficulties in isolating drug metabolites from biological materials lie in their low concentrations in urine, bile or blood, and in the high content of endogenous compounds and food constituents. The physico-chemical properties of the more strongly polar metabolites further complicate their isolation in pure form. Preparative liquid chromatography ('prep-LC') offers many advantages. As is now exemplified, it enables the products to be isolated in quantities sufficient for full structure elucidation by spectroscopic and chemical methods. It can be applied to complex mixtures of metabolites of widely varying polarities. Since separation is done at low temperature, normally no stability problems arise. Recoveries can easily be checked if radioactively labelled drugs are used.

In practice, a two-step procedure has proved useful. A pre-purification is done to remove the bulk of endogenous substances, e.g. inorganic salts, urea, amino acids, from the biological material; this is normally achievable by classical column chromatography (coarse-grade packing material, low pressure) with adsorbents such as Amberlite XAD-2 resin and simple technical equipment. Then, depending on polarity, the individual metabolites are separated by high-resolution prep-LC in a normal or reverse phase mode (NP, RP) as appropriate. Gradient elution techniques can greatly improve column performance, but require sophisticated equipment.

While most drugs are more or less lipid-soluble, favouring rapid absorption, distribution and transport to the target organ, their renal and biliary elimination necessitates enzymic conversion into Phase I and Phase II products of enhanced polarity and water solubi-

lity (Fig. 1).   To establish the pathways, metabolites have to be
separated out for structure elucidation. Besides their low concentration amidst large amounts of endogenous and food constituents, there
is the problem that individual metabolites frequently possess closely
related chemical structures which impede their separation.   Moreover,
the physico-chemical properties of the more polar metabolites, notably
conjugates, are unconducive to their isolation (Fig. 1).   Wherever
possible, radiolabelled parent drug should be used in metabolism
studies, since it enables metabolites to be distinguished amongst
naturally occurring constituents and also to be quantitated. Chromatography is the classical means for separating and isolating drugs and
their metabolites.   The most advantageous approach, in our experience,
is prep-LC.

CHROMATOGRAPHIC SEPARATION AND ISOLATION APPROACHES FOR METABOLITES

     Table 1 gives various possible approaches together with advan-
tages and disadvantages .which need little amplification.   Although
CCD [1] and its more modern version, DCCC [2], have drawbacks as indicated, advantages include avoidance of autoxidation because a closed

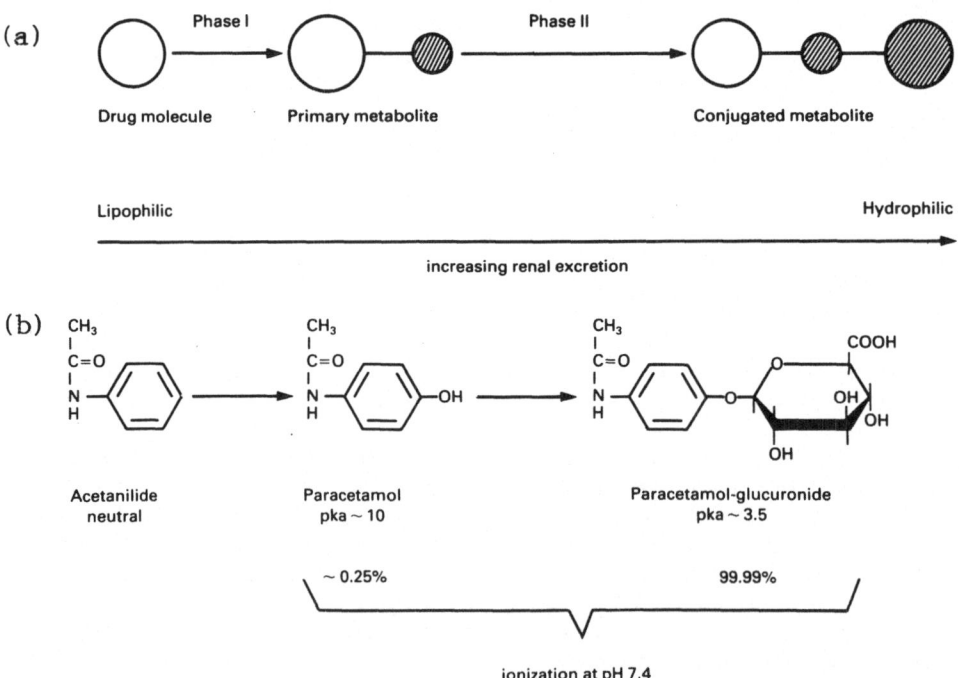

Fig. 1. Drug biotransformation. (a) Principles. (b) Effect on
polarity, exemplified by acetanilide: its weakly acidic product,
paracetamol, gives a glucuronide conjugate which is completely
ionized and highly water-soluble [3]. [Phase I may be by-passed.–Ed.]

Table 1. Ways to isolate and separate drug metabolites for structure elucidation.

Method	Advantage	Disadvantage
Counter-current distribution (CCD) and Droplet counter-current chromatography (DCCC)	High capacity (~1-4 g) Complete recovery, since no solid support that could adsorb Non-destructive Radiotracing feasible	Low resolution Direct detection not feasible Limited applicability Slow
Thin-layer chromatography (TLC) and Paper chromatography (PC)	Versatile application Fast; simple operation Functional groups identifiable by sprays Radiotracing feasible	Low capacity Autoxidation Incomplete recovery, especially of polar metabolites
Gas chromatography-Mass spectrometry (GC-MS)	On-line elucidation of structure High resolution	Low capacity Derivatization necessary No recovery control Limited applicability Complicated operation
Preparative liquid chromatography (prep-LC)	High capacity and resolution Versatile application Almost complete recovery Non-destructive Radiotracing feasible	Complicated operation Costly solvents and column packings

apparatus is used, and they do allow preparative-scale separation of polar compounds. Thus, CCD has been applied to the isolation of sulfinpyrazone metabolites [4].

TLC is widely used for separating and isolating drug metabolites (as reviewed [5]; see also refs. in #C-1). Despite considerable disadvantages, it has advantages including the possibility of radio-scanning of the chromatograms. With GC-MS direct radiotracing is precluded, but this drawback can be partly overcome by stable isotope labelling [6]. However, positive structure elucidation is usually not obtainable merely from the MS, and many drug metabolites, especially if polar, cannot be investigated without chemical modification or thermal degradation, if at all. Since GC peaks of interest will include contaminants, it is very hard to decide whether the MS of a given GC peak is that of a drug metabolite or not. Amongst numerous papers on metabolite separation and identification by GC-MS, a recent example [7] concerns alprenolol metabolites in urine and bile, with or without enzymic hydrolysis.

Table 2.   Approaches to prep-LC isolation of drug metabolites.

LC mode		Packings	Advantage	Disadvantage
Steric exclusion	GFC	Dextran gel (Sephadex)	Simple operation	Low resolution Adsorption effects
	GPC	Crosslinked polystyrene, polyamide		
Liquid-solid (LSC)		Porous silica	High selectivity High resolution	Limited to lipophilic compounds
		Alumina		
Liquid-liquid (LLC)	Normal phase (NP)	Porous silica/H_2O Bonded phases (diol, NH_2)	High selectivity High resolution Versatile application	Limited to pH $\leq$ 8 Non-rigid
	Reverse phase (RP)	Bonded phases (C-2 – C-18, CN) Styrene-divinylbenzene copolymer (Amberlite XAD-resins)		
Ion exchange (IEC)	Cation Anion	Cross-linked polystyrene – sulphonic acid – quaternary amine	Selective for charged molecules	Adsorption effects Use of non-volatile buffers

EDITORIAL NOTE ON NOMENCLATURE.- The term 'steric exclusion' to cover gel filtration/permeation chromatography has been introduced editorially. For bonded-phase packings, the validity of the designation 'liquid-liquid' is arguable (see Preface), as is the assignment of the CN type to reverse rather than normal phase. Moreover, the general context is HPLC, notwithstanding the authors' adoption of the term 'LC' for all approaches.

PREPARATIVE LIQUID-CHROMATOGRAPHIC METHODS FOR METABOLITES

Table 2 summarizes the four basic LC methods, with different mechanisms or processes, and their advantages and disadvantages. In GFC (for aqueous solvents) and GPC (for organic solvents), where small molecules permeate most of the pores and are retained, powerful separations are obtainable, sometimes helped rather than hindered by adsorption effects that enable compounds of the same molecular size to be separated. Since the gels are semi-rigid or non-rigid, their use in modern prep-LC under high pressure is restricted. In drug metabolism studies, preparative gel chromatography is applied mainly as an initial group-fractionating step for conjugates [8,9].

Adsorption chromatography (LSC) involves particles, e.g. porous silica or alumina, with a large area of surface, on which sample molecules are retained by affinity. Because of irreversible adsorption of polar compounds onto the hydrophilic packing material, LSC is generally unsuitable for separating and isolating drug metabolites, especially conjugates. In LLC (partition chromatography), with a stationary phase differing in composition from the moving liquid phase, sample molecules distribute between the two phases as in liquid-liquid extraction. The two phase liquids may be immiscible, or the stationary-phase liquid may be chemically bonded to the support. In normal-phase (NP) LLC, stationary coating liquids are generally polar, e.g. water or ethylene glycol, and the mobile phase non-polar, e.g. heptane or dichloroethane. In reverse-phase (RP) LLC the stationary phase is non-polar, e.g. hydrocarbons (Table 2), and water/alcohol mixtures are used as mobile phases. Merits of the two approaches (Table 2) include applicability to diverse sample types, polar and non-polar, through the large number of partitioning phases that are available, as exemplified below for drug metabolites.

IEC (Table 2) gives good separation of charged compounds, exchanging with the counter-ion of the solid electrolyte (examples in [10]); but its application in drug metabolism studies is limited by irreversible adsorption on the matrix and buffer involatility.

DEFINITION OF PREPARATIVE LC AND THEORETICAL CONSIDERATIONS

According to Verzele & Geeraert [11], LC is termed 'preparative' when it isolates a fraction, pure or not so pure, from a mixture for further use (for spectroscopy, identification, synthesis, commercial purposes, etc.). The choice between the two general approaches depends on how much product is needed [12], as shown in Fig. 2. For sample weights < 1 mg/g of adsorbent, i.e. the analytical range, peak retention (k' values) and column efficiency change little with sample size. Mere scaling-up, with large diameter columns, allows mg amounts to be isolated, with analytical-scale resolution, and is thus apt for preparing modest amounts of highly purified compounds from complex mixtures, e.g. drug metabolites from biological media. In the second preparative approach, large diameter columns are operated in an 'overload' fashion with sample loads >1 mg/g of packing material. Overload has been suggested to occur when either the column efficiency (plate number) decreases, or the peak width increases, or the capacity factor is lowered by ≥10% with increasing sample size [11]. Overload prep-LC serves mainly to prepare one or two purified products in large amount, with relatively short separation times because high flow rates can be used without seriously impairing resolution.

Optimization of conditions for prep-LC proceeds similarly to that for analytical separations. It is necessary to use:

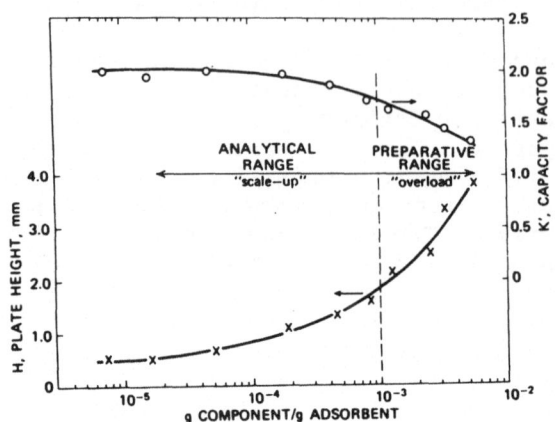

	"Scale-up" separation	Column "overload" separation
Column diameter	Large, up to 5 cm	Large, up to 5 cm
Sample weight	< 1 mg/g of adsorbent	> 1 mg/g of adsorbent
Resolution power	High; little change of peak retention (k'-values) with sample size	Low; k'-values are decreased > 10% by increasing sample size
Column efficiency	Independent of sample size as in analytical separations	Dependent on sample size

Fig. 2. Approaches in preparative liquid-chromatographic isolation.

(i) the proper solvent strength, expressed by the capacity factor k'

$$k' = \frac{t_R - t_0}{t_0}$$

(ii) good separation selectivity α

$$\alpha = \frac{k'_2}{k'_1} = \frac{t_{R_2} - t_0}{t_{R_1} - t_0}$$

(iii) adequate column efficiency N

$$N = 16\left(\frac{t_R}{w}\right)^2$$

The effect of these chromatographic parameters on analytical and scaled-up (non-overload) separation can be predicted by the resolution (R_S) relationship,

$$R_S = \frac{1}{4} \times \sqrt{N} \times \frac{(\alpha - 1)}{\alpha} \times \left(\frac{k'}{1 + k'}\right)$$

because α and k' are essentially independent of mobile phase velocity and other kinetic effects.

Since this article is aimed rather at the practitioner faced with separation and isolation problems in drug metabolism, the more theoretically interested reader should consult recent excellent reviews [e.g. 11-13] or books.

EQUIPMENT AND MATERIALS IN PREPARATIVE LIQUID CHROMATOGRAPHY

Two firms at present market apparatus (Jobin-Yvon, France - Chromatospac, Miniprep; Waters, U.S.A. - Prep LC/500 System), as fully described in their brochures and literature cited therein and in a review [14]. The apparatus is constructed especially for the isolation of large sample amounts with short, wide columns in an overload mode, and its use in drug-metabolism studies with highly contaminated samples seems rather inappropriate in our experience.

Instrumentation

The pump that is the heart of the system can be reciprocating (e.g. Altex 110A) or of pneumatic amplifier type (e.g. Haskel). The solvent delivery should be fast, at least 10 ml/min. Since columns of various sizes and packing materials of different permeabilities are used, the pump should be capable of a relatively high pressure (~200 bar).

Gradient elution is of great advantage in isolating drug metabolites, and can be achieved by two different set-ups: those that mix solvents at atmospheric pressure and then pump the mixture at high pressure into the column [cf. #E-5 in Vol. 5 and #D-5 in Vol. 7, this series –*Editor*], and those where the respective solvents are pumped at high pressure into a mixing chamber preceding the column. In the former, the separately stored solvents (up to 3) on the intake side of the single pump are combined via a microprocessor-controlled proportioning solenoid valve (e.g. Spectra-Physics, SP 8700). In the latter there are two pumps with programmed delivery (e.g. Kontron or Waters Gradient system). Either mode is versatile, and can help in drug-metabolism studies as long as a chosen gradient profile is reproducible.

In prep-LC, larger sample aliquots can be conveniently introduced into the column with a 6-port loop *injector* of large volume (up to 10 ml; e.g. Valco or Rheodyne). The sample should be loaded over the entire column cross-section, since this permits better use of the total packing with reduced column overload. The volume that can be injected will depend on the column i.d., sample solubility, and the mobile/stationary phase combination. In a non-overload column condition, which we prefer for metabolite isolation, this volume generally should not exceed ~one-third of the volume of the earliest peak of interest [12].

Preparative columns are made either of stainless steel (e.g.

Table 3. Packings and columns for prep-LC of drug metabolites.

Column sizes: glass (heavy-wall), 1.25-5 cm X 30-100 cm;   stainless steel, 0.4-1 cm X 25 cm;   pre-columns, 0.4-1 cm X 2-5 cm.

Purpose	Suitable packing	Column
Pre-purification	Coarse-grade styrene-divinylben-zene copolymer (Amberlite XAD-2 resin, 35-50 mesh)	Glass, large i.d.
Separation and isolation of lipophilic metabolites	Porous silica gel + gypsum (TLC material), 10-40 µm	Glass, large i.d.
	ditto (Lichroprep), 25-40 µm	ditto
	Porous silica gel (Lichrosorb), 10 µm	Glass, steel; small i.d.
Separation and isolation of polar and lipophilic metabolites	Bonded phase (RP2....RP18; Lichro-prep), 25-40 µm	Glass, large i.d.
	ditto (ditto; Lichrosorb), 10 µm	Glass, steel; small i.d.
	Micronized Amberlite XAD-2 or XAD-4 resin, 12 µm	Glass, small and large i.d.

Altex), which can be operated up to ~250 bar pressure, or of heavy-wall glass (e.g. Chromatronix) for below ~15 bar. The i.d. range used in our laboratory is 0.4-5 cm (Table 3).

Spectrophotometric detectors do not necessarily have to be of high sensitivity since solute concentrations are generally high in prep-LC. UV-detectors with variable wavelength (e.g. Uvicon 725, Kontron) offer the advantage of 'de-tuning' the wavelength from the solute absorption maximum, in order to decrease the sensitivity of the monitoring.   Alternatively, a stream splitter on the column exit can supply only a small fraction of the effluent to the detect-or. If radiolabelled parent drug is used, a *radioactivity monitor* (RAM; e.g. Berthold LB 503) is the first choice for detecting the individual metabolites without interference by endogenous compounds. For preparative work, RAM flow-cells should be filled with a solid scintillator to avoid contamination of the separated fractions. Moni-toring with liquid scintillator in a flow-cell is feasible by stream splitting of the effluent and by having a separate pump to mix it with a suitable liquid scintillator [15].   [Cf. #NC(A)-4.-*Ed.*] With the liquid-scintillation counting version there may be benefit to sensitivity, but 10-15% loss of the separated material. Simple off-line radiometry of aliquots from collected fractions is, of course, satisfactory too. Automatic *collectors* (e.g. LKB 2111 Multifrac) that allow small or large fraction volumes to be collected are convenient

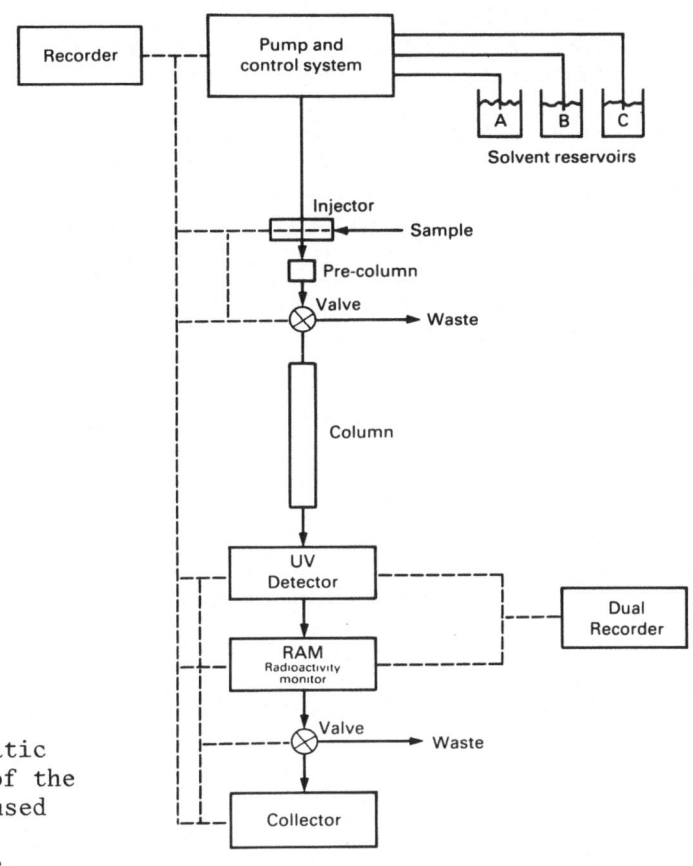

Scheme 1. Schematic
representation of the
prep-LC system used
in isolation of
drug metabolites.

in drug-metabolite separation and isolation, where a large number of
compounds usually have to be collected in one run.

With a modern LC apparatus, equippped with a microprocessor, it
is nowadays possible to control not only pump functions and gradient
profiles, but also the injection system, the fraction collector and
valves as depicted in Scheme 1 (e.g. Spectra Physics SP 8000).   This
automatic system can be used to repeat chromatographic runs in narrow
columns so as to get enough material for structure elucidation [16].

Column packings for prep-LC of drug metabolites

With increased column i.d., as the usual means in prep-LC of increasing sample capacity, generally there is an increase in plate
number, i.e. efficiency. The packing inhomogeneities that occur in
narrow columns due to wall effects and seize-up of particles in the
column bed appear to be less significant in wider columns [17]. Table
3 shows appropriate packings and column types for drug-metabolism

studies. LLC in the NP or RP mode (Table 2) is the preferred means, the advantages including good recovery and the use of mobile phases in which the samples can easily be dissolved.

As packing materials, commercially available (Table 3), we recommend porous silica gel which may have had gypsum added (as with TLC silica gel G, or Lichroprep Si 60), or RP-2, -8 or -18 material, of 10-40 μm particle size depending on the column i.d. A supplementary RP material is a micronized XAD resin (Table 3): the mobile phase is polar, but contains a small proportion of a lipophilic solvent that is preferentially taken up by the resin and thus forms a stationary phase *in situ* [18].

At the outset of the isolation, glass columns of large i.d. packed with coarse XAD-2 achieve concentration and pre-purification.

Column packing procedures in prep-LC

For the above-mentioned columns, wet filling techniques are used exclusively. Amongst the 4 procedures, the choice depends on the column i.d., the packing material and the particle size.

Slurry-sedimentation method: for packing coarse-grade XAD-2 (in water; degassed) in wide glass columns that are used in the classical chromatographic mode. The bed is formed merely by filtration as the solvent is allowed to drain from the column outlet during the sedimentation process.

Slurry-gypsum method: this is based on a published technique [19]. Our own hitherto unpublished version employs silica gel G or 10 μm Lichrosorb Si 60 + 15% gypsum, suspended in distilled water at 60°, 3.5 ml/g (round-bottomed flask; 4.5 g of adsorbent per 10 ml of column void volume). After degassing for 15 min at 60°, (water-bath; rotary evaporator, reduced pressure), the hot suspension is poured into the heated column (1.27 × 30 cm up to 5 × 50 cm, Chromatronix) which is connected with a second column as reservoir. The column assembly is completely filled with water at 60° and immediately pressurized with nitrogen (7-20 bar) via the top plunger, the water being allowed to flow out through the bottom plunger. During the filling process the columns are mounted vertically in the set-up shown in Plate 1, which allows an oscillating radial movement of the column assembly (one-third full rotation, ~60 oscillations/min). Columns of i.d. 1.27 cm can be packed without movement. The pressure is released before the bed surface has become dry. The column is left overnight at room temperature. The adsorbent is activated by percolating ethanol through the column, subsequently displaced by the chosen eluent.

Low pressure slurry method: used with glass columns for the above-mentioned RP packings. Essentially it follows the slurry-

Plate 1. Apparatus
for packing prepar-
ative columns by the
slurry-gypsum method.

gypsum procedure, but it is performed at ambient temperature and without movement of the column assembly [18].

   High pressure slurry method: the non-balanced density technique
used in analytical HPLC [20, 21], performed with 10 µm packings (porous silica or bonded phase) in steel columns of 0.4-1 cm i.d. (Altex).
Briefly, a slurry of silica gel in carbon tetrachloride or RP material
in methanol (6-20%, w/v) is forced from a larger diameter reservoir
(volume 300 ml) with heptane or methanol respectively, into the column
at a very high linear velocity, using the apparatus depicted in Plate
2. A constant pressure Haskel pneumatic amplifier pump capable of an
output pressure of 300 bar serves as the pressure source.

SOLVENT SYSTEMS FOR PREPARATIVE CHROMATOGRAPHY OF DRUG METABOLITES

   Proper choice of solvents is vital for success. In the usual

Plate 2. Apparatus
for packing prepar-
ative columns by the
high pressure slurry
method.

way, column selectivity α and capacity factor k' (solvent strength)
are optimized by selecting the composition of the mobile phase in
test runs. TLC and HPLC can often be used for this purpose. Solvent
strength is first adjusted in the right range; then, if a separation
is not accomplished, solvents of similar strength but consisting of
different compounds are tested to optimize α.

     However, other solvent properties must be considered too. Mobile
phases should be easily volatilizable to guarantee simple and mild work-
up of the eluted fractions. A special problem in prep-LC is sample
solubility. The sample has to be dissolved in an adequate volume of
the mobile phase. Injection in a stronger solvent can cause severe
impairment of column resolution. Mobile phases of low viscosity are
preferable, to optimize resolution and avoid excessive pressure.
Single-phase mixtures, well below the saturation point, are best, so
that performance is not impaired by the usual variations in ambient

temperature. For mixtures of metabolites, differing widely in polarity, gradient elution is very effective; hence the solvents should be
suitable for gradient mixing. Moreover, solvents must be highly pure
since impurities are concentrated in the individual, evaporated fractions, and the resulting contamination can impede spectroscopic
structure elucidation. Finally, solvent prices should be minimal,
since large volumes are needed in prep-LC.

Metabolites can be well separated by NP- and RP-LC with simple
and essentially similarly composed solvent systems. In the NP mode
with silica gel, single-phase ternary solvent mixtures are very useful. A lipophilic solvent, e.g. 1,2-dichloroethane, and a lower aliphatic alcohol, e.g. ethanol, as major components are mixed with a
minor amount of water or diluted acid or base, as in Fig. 3. When the
column is equilibrated, the polar components of the mixture build up
a stationary liquid phase on the hydrophilic silica gel, and solute
molecules distribute between the two liquid phases according to their

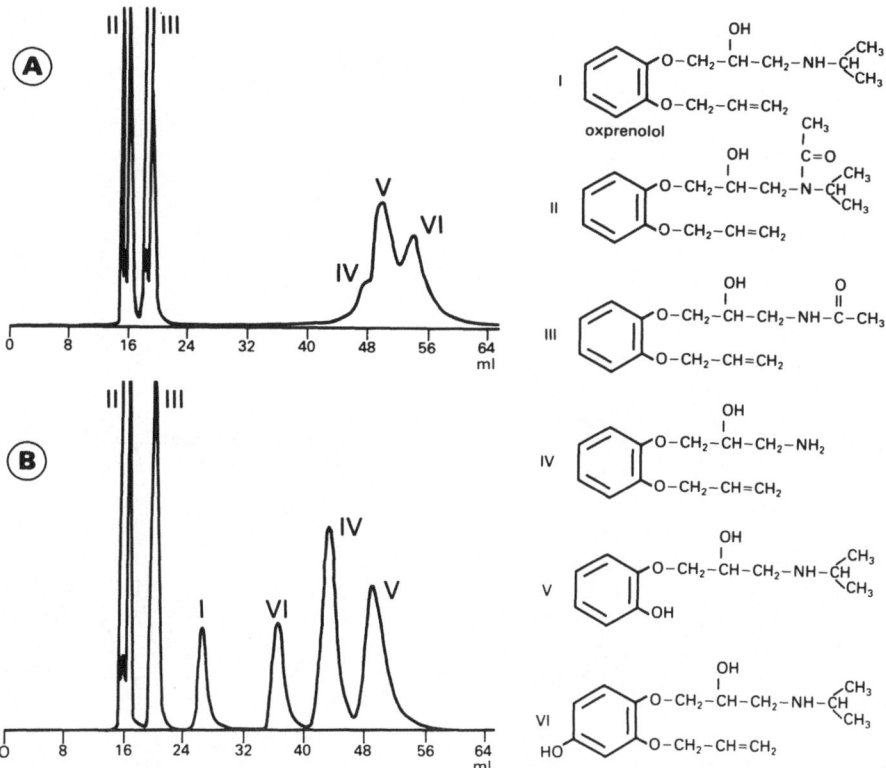

Fig. 3. Solvent selectivity in liquid-liquid chromatography.
Column: 1 x 25 cm (steel), 10 μm Lichrosorb Si 60. Flow rate 4 ml/min.
Detection at 270 nm. Injection: 0.3 mg of each pure compound in 200 μl.
Mobile phase: A, 1,2-dichloroethane-methanol-aq. NH_3 (80:20:0.5);
               B, 1,2-dichloroethane-isopropanol-aq. NH_3 (55:43:2).

partition coefficients.   Account must also be taken of adsorption of solute onto the polar functions of the gel.   Selectivity can be varied widely by merely changing the polar components of the mobile phase, as illustrated in Fig. 3 where incomplete separation of the oxprenolol derivatives and metabolites as in **A** is remedied merely by changing the alcohol component (see **B**).

For RP-LC the commonest solvent is water mixed with methanol or acetonitrile. Silica-based bonded phases can be used only in the pH range ~2-8, because of stability problems, whereas XAD-2 can be used over the entire pH range, suitably with ternary mixtures of water, alcohol and a non-polar solvent (toluene or dichloroethane). As indicated above, the RP retention mechanism is as for NP, but in a reciprocal sense in respect of stationary and mobile phase [18, 22].

EXAMPLES OF APPLICATIONS IN DRUG-METABOLISM STUDIES

For many years, prep-LC has been used effectively in our laboratories for separating and isolating metabolites in amounts adequate for full structure elucidation.   This is now illustrated with some examples, especially in the partition mode.   Before any prep-LC separation of a mixture is started, the metabolite pattern   should be ascertained.   Besides TLC, e.g. combined with autoradiography for ^{14}C-labelled drugs [23], RP gradient HPLC is the method of first choice, enabling native biological samples to be examined.   After

Fig. 4. Characterization of metabolites of ^{14}C-oxaprotiline in urine collected for 5 or (dog) 4 days after oral dosage (man, 50 mg; dog, 20 mg/kg; rat, 50 mg /kg).   Recoveries of [14]C = 102%, 98% and 105% respectively.
Column, 0.46 × 25 cm (steel), 10 µm Lichrosorb RP 8; mobile phase: linear gradient composed of 0.01 M HCl (A), ethanol (B) and acetonitrile (C), 0-30 min with 5→35% B, C held at 5%; 30-55 min 35→90% B, C held at 5%. Flow rate 2 ml/min. Effluent subjected to radiomonitoring.
Injection: 5 ml of unprocessed urine from each species, preconcentrated by a pre-column.

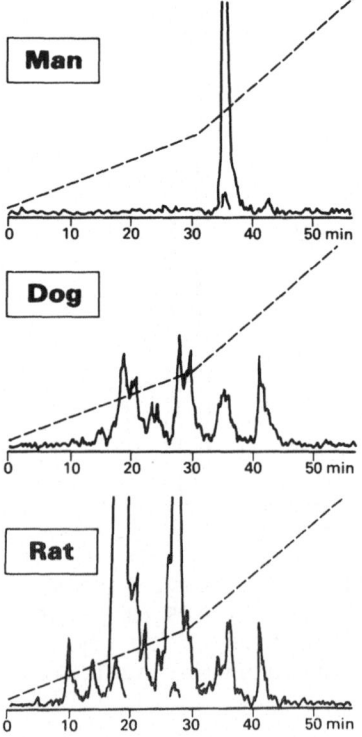

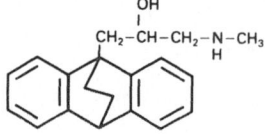

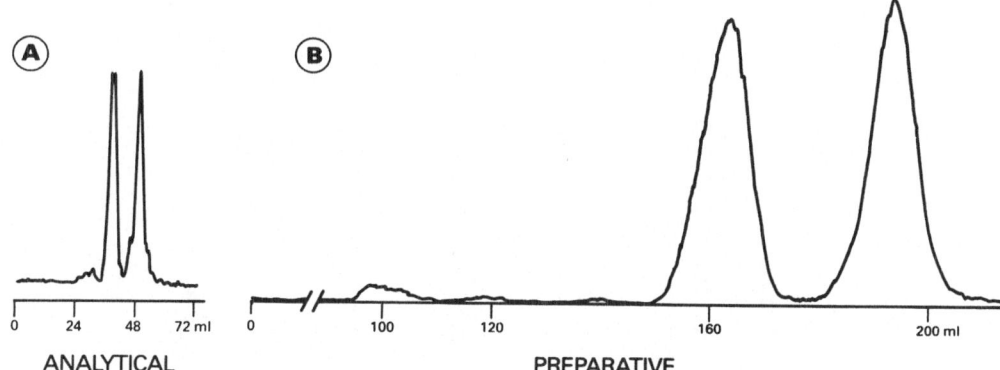

Fig. 5. Preparative separation of diastereoisomeric glucuronides of ^{14}C–oxaprotiline.
Columns (glass): **A**, 1.27 × 30 cm; **B**, 2.54
× 50 cm; 12 μm XAD-2.
Mobile phase: methanol–water–acetic acid–
1,2–dichloroethane (60:38:1:1). Flow rate
2 ml/min. Effluent radiomonitored.
A, analytical run (human urine).
B, preparative run; injection of 16 mg [14]C–
metabolites (polar) in 750 mg crude material
(pre-concentrated from human urine).

enrichment on a pre-column, all metabolites can be eluted by means of
a gradient, irrespective of their physico-chemical properties [25].
Thus, in Fig. 4 the urinary metabolite pattern in man, dog and rat,
following administration of ^{14}C–oxaprotiline, is recorded by analytical RP gradient HPLC (our unpublished work).

Prep-LC separation is likewise exemplified by this drug [Fig. 5].
Since it has a chiral centre, direct conjugation with glucuronic acid
leads to a pair of diastereoisomeric glucuronides (our unpublished
work). Chromatogram **A** is their analytical separation from urine with
a micronized XAD-2 column and isocratic elution. Chromatogram **B**
shows that equivalent resolution was attained when a prep-LC column
was loaded with 16 mg of polar ^{14}C–metabolites in 750 mg of crude
material. This illustrates the large capacity and high selectivity
of the preparative system, even when operated in an overload condition,
with regard to the impurities.

For ^{14}C–carbamazepine metabolites in rat bile, Fig. 6 shows
preparative separation of the more lipophilic fraction by NP–LC on
silica gel. Good resolution of chemically closely related compounds
is again evident. Thus, the peak eluted with ~80 ml and the last two
peaks (120–130 ml) represent three metabolites, hydroxylated in one
of the two phenyl rings, viz. 1–OH–, 4–OH– and 3–OH–carbamazepine
respectively (unpublished work). Fig. 7 shows the gross separation
of the very complex metabolite mixture of the hydrophilic non–extractable fraction, essentially glucuronides and sulphates, by RP–HPLC.

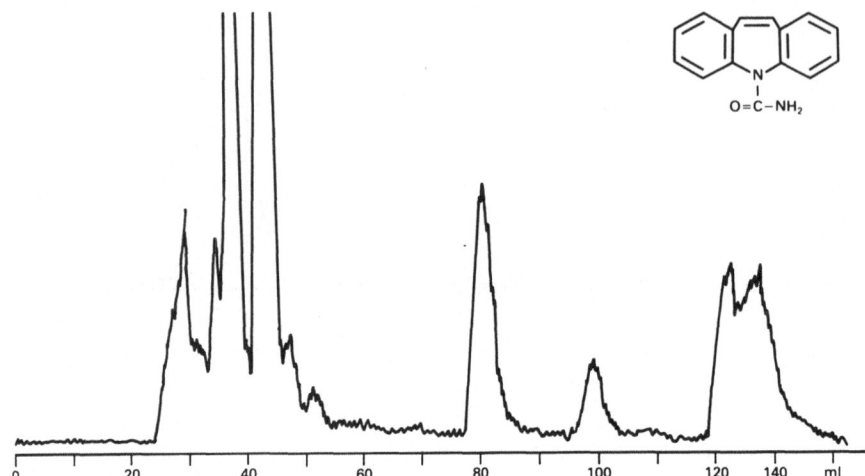

Fig. 6. Preparative separation of ^{14}C-carbamazepine metabolites.
Column: 1.27 x 30 cm (glass). 10 μm Lichrosorb Si 60; mobile phase:
1,2-dichloroethane-methanol-aq. NH$_3$ (90:10:0.5). Flow rate 1 ml/
min. Effluent radiomonitored.
Injection: 4 mg ^{14}C-metabolites (lipophilic) in 100 mg crude
material (extracted from rat bile).

It is evident that,with a gradient, more than 1 g of crude material
was successfully separated and the individual metabolites were considerably enriched.

   The use of silica gel G with its wide range of particle size
is exemplified in Fig. 8. The extractable, lipophilic ^{14}C-metabol-
ites of clomipramine from human urine were separated with a simple
ternary mixture. From the peak eluting with ~330 ml, for instance,
3 isomeric monohydroxylated metabolites were isolated by further
chromatography. Corresponding hydroxylated desmethylclomipramine
metabolites were obtained from the last peak (~720 ml)(unpublished
work along with H. Mory).

   As a final example, the usefulness of relatively narrow, high-
resolution columns is depicted in Fig. 9. Urinary polar metabolites
of ^{14}C-oxcarbazepine were pre-concentrated with coarse-grade XAD-2
and run with 10 μm Lichrosorb RP. Though more than 800 mg of crude
material was injected, highly enriched metabolite fractions were
obtained by gradient elution. After derivatization, they were
directly applied to structure elucidation with spectroscopic methods
and were recognized as closely related glucuronides (H. Schütz,
personal communication, 1981).

   These examples prove the great advantage of preparative parti-

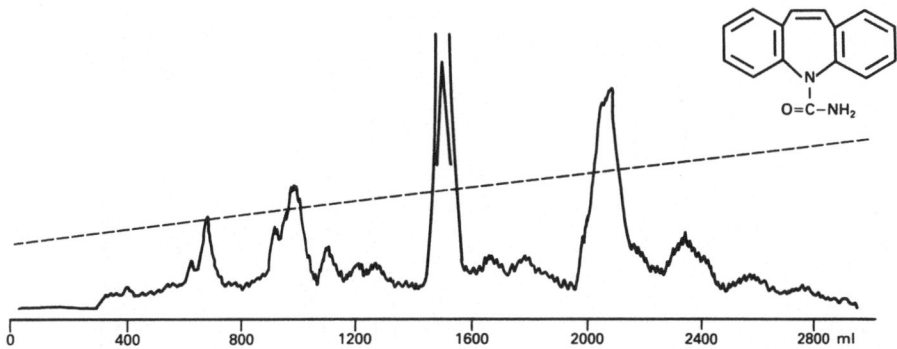

Fig. 7. Preparative separation of ^{14}C–carbamazepine metabolites.
Column: 5 × 30 cm (glass), 12 μm XAD–2; mobile phase: linear
gradient composed of 2 l of methanol–water–1,2–dichloroethane–
acetic acid (75:73:1:1) (A) and 2 l of the same (80:18:1:1) (B).
Flow rate 2 ml/min. Effluent radiomonitored.
Injection: 54 mg ^{14}C–metabolites (polar) in 1.1 g crude material
(pre–concentrated from rat bile).   *(Unpublished experiment.)*

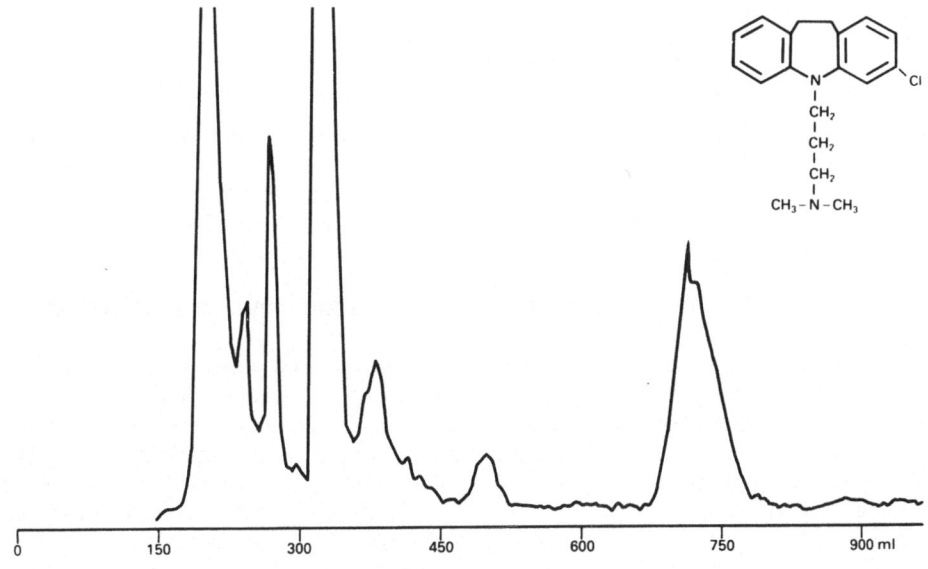

Fig. 8. Preparative separation of ^{14}C–clomipramine metabolites.
Column: 2.54 × 50 cm (glass), 10–40 μm silica gel G; mobile phase:
n–propanol–1,2–dichloroethane–aq. NH$_3$ (100:100:2). Flow rate
2 ml/min. Effluent radiomonitored.
Injection: 15 mg ^{14}C–metabolites (lipophilic) in 350 mg crude
material (extracted from human urine).

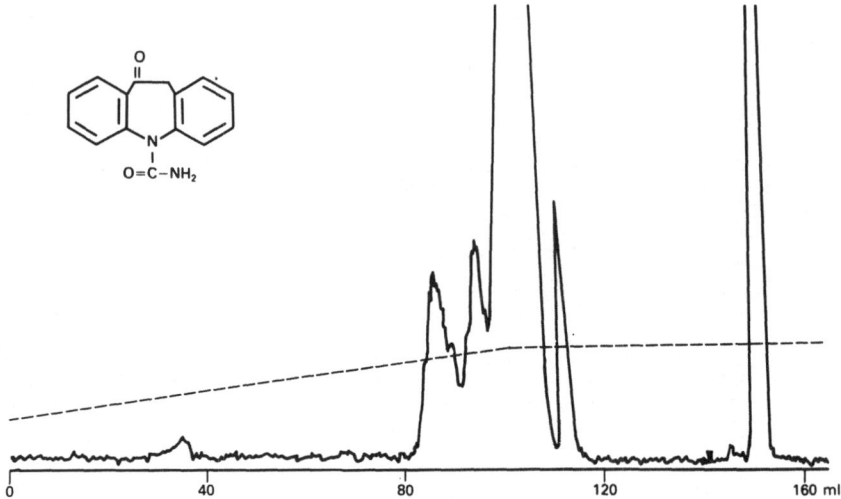

Fig. 9. Preparative separation of ^{14}C-oxcarbazepine metabolites.
Column: 1 x 25 cm (steel), 10 µm Lichrosorb RP-18; mobile phase:
linear gradient composed of water (A) and methanol (B): 0-50 min,
10→30% B; 50-80 min, isocratic with 30% B. Flow rate 2 ml/min.
Effluent radiomonitored.
Injection: 18 mg ^{14}C-metabolites (polar) in 800 mg crude material
(pre-concentrated from human urine).

tion chromatography, NP and RP, for separating and isolating drug
metabolites. The methods are characterized by a large capacity, a
high resolving power, a high degree of selectivity, and versatile
application for polar and non-polar metabolites in complex and very
impure mixtures.

GENERAL PROCEDURE FOR SEPARATING AND ISOLATING DRUG METABOLITES

Although obviously no exact and universally applicable prescrip-
tion can be given, our experience of prep-LC techniques and of the
chromatographic properties of metabolites allows a general approach
to be recommended. Scheme 2 indicates how drug metabolism studies
are usually approached in our laboratories. Provided that radiolabelled parent drug is used, the biological sample containing the
metabolite mixture is firstly characterized by analytical RP gradient HPLC. This technique also serves in the further course of
separation and isolation as a qualitative and quantitative control.
Pre-purification is next performed to substantially remove endogenous substances, normally achievable by classical column chromatography with adsorbents such as coarse-grade XAD-2 resin. Then the
concentrate is partitioned between an aqueous and an organic phase to
give a polar and a non-polar fraction, and each subjected to prep-LC
as discussed above.

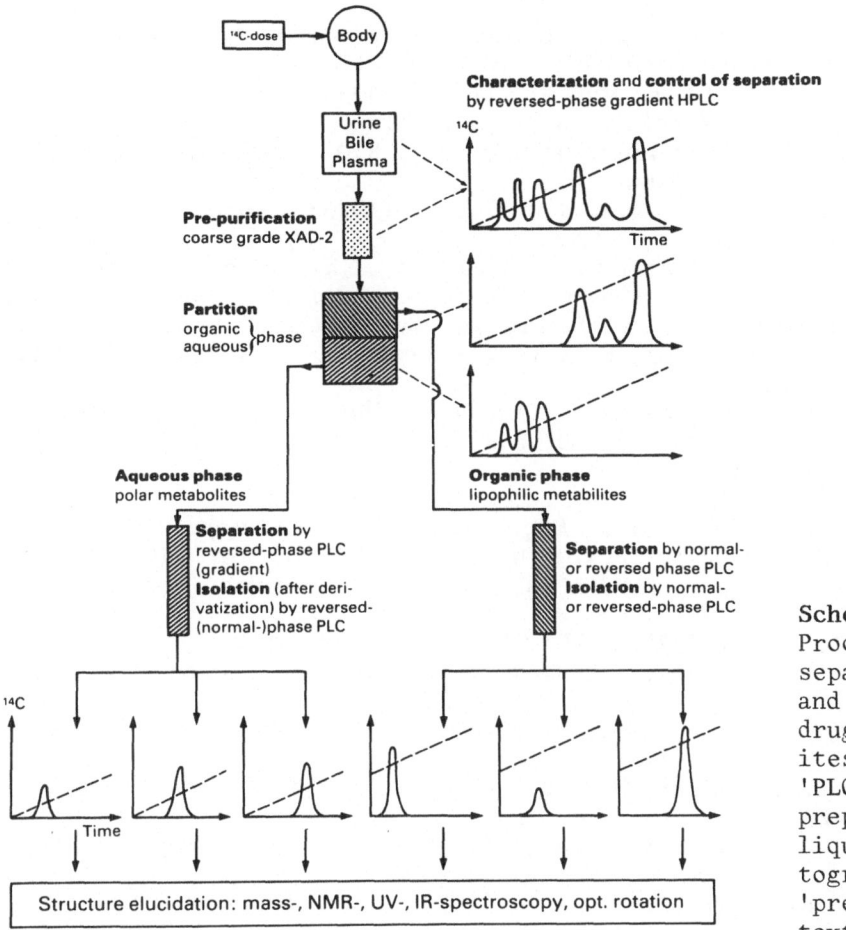

Scheme 2.
Procedure for
separating
and isolating
drug metabol-
ites.
'PLC' denotes
preparative
liquid chroma-
tography (cf.
'prep-LC' in
text).

Gradient elution techniques can often greatly improve chromatographic performance. If necessary, derivatization steps are included, e.g. for glucuronides. Finally structure elucidation (as
in Scheme 2) is performed on the pure individual metabolites.

References

1. Craig, L.C. (1960) in *Analytical Methods of Protein Chemistry*,
    Vol. 1 (Alexander, R. & Block, R.J., eds.), Pergamon, Oxford,
    pp. 121–160.
2. Tanimura, T., Pisano, J.J., Ito, Y. & Bowman, R.L. (1970)
    Science 169, 54–56.
3. Williams, R.T. (1971) in *Concepts in Biochemical Pharma-
    cology*, Part 2 (Brodie, B.B. & Gillette, J.R., eds.) [Vol. 28/2
    of *Handbook of Experimental Pharmacology*], Springer, Berlin,
    pp. 226–242.

4. Dayton, P.G., Sicam, L.E., Landrau, M. & Burns, J.J. (1961) *J. Pharmacol. Exp. Ther. 132*, 287–290.

5. Titus, E.O. (1971) as for 3., pp. 142–151.

6. Prox, A. (1973) *Xenobiotica 3*, 473–492.

7. Hoffmann, K.-J., Arfwidsson, A., Borg, K.O. & Skånberg, J. (1979) *Xenobiotica 9*, 93–106.

8. Radomski, J.L., Rey, A.A. & Brill, E. (1973) *Cancer Res. 33*, 1284–1289.

9. Richter, W.J., Kriemler, P. & Faigle, J.W. (1978) in *Recent Developments in Mass Spectrometry in Biochemistry and Medicine*, Vol. 1 (Frigerio, A., ed.), Plenum, New York, pp. 1–14.

10. Arndts, D. & Rominger, K.L. (1978) *Arzneim.-Forsch./Drug Res. 28*, 1951–1960.

11. Verzele, M. & Geeraert, E. (1980) *J. Chromatog. Sci. 18*, 559–570.

12. Di Stefano, J.J. & Kirkland, J.J. (1975) *Anal. Chem. 47*, 1103A–1108A & 1193A–1204A.

13. Wehrli, A. (1975) *Z. Anal. Chem. 227*, 289–302.

14. Nettleton, D.E. (1981) *J. Liq. Chromatog. 4 (Suppl. 1)*, 141–173.

15. Dugger, H.A. & Orwig, B.A. (1980) *Drug Metab. Rev. 10*, 247–269.

16. Hupe, K.-P., Lauer, H.H. & Zech, K. (1980) *Chromatographia 13*, 1–8.

17. Wolf III, J.P. (1973) *Anal. Chem. 45*, 1248–1250.

18. Dieterle, W., Faigle, J.W. & Mory, H. (1979) *J. Chromatog. 168*, 27–34.

19. Sie, S.T. & van den Hoed, N. (1969) *J. Chromatog. Sci. 7*, 257–266.

20. Linder, H.R., Keller, H.P. & Frei, R.W. (1976) *J. Chromatog. Sci. 14*, 234–239.

21. Coq, B., Gonnet, C. & Rocca, J.L. (1975) *J. Chromatog. 108*, 249–262.

22. Brown, P.R. & Krstulovic, A.M. (1979) *Anal. Biochem. 99*, 1–21.

23. Dieterle, W., Faigle, J.W., Montigel, C., Sulc, M. & Theobald, W. (1977) *Eur. J. Clin. Pharmacol. 11*, 367–375.

24. von Hodenberg, A., Klemisch, W. & Vollmer, K.-O. (1977) *Arzneim.-Forsch./Drug Res. 27*, 508–511.

25. Dieterle, W., Faigle, J.W. & Moppert, J. (1980) *Arzneim.-Forsch./Drug Res. 30*, 989–993.

#A-3

PROBLEMS ARISING FROM ASSAYING LOW CONCENTRATIONS BY HPLC

C.R. Jones

Wellcome Research Laboratories
Beckenham, Kent BR3 3BS, U.K.

Assays of drugs and metabolites by HPLC are frequently carried out at very high sensitivity, i.e. conditions of low signal-to-noise ratio. This brings in train a number of problems; some are well known, others are not suspected by those workers who deal only with samples at high concentration. The examples given represent problems that have occurred in a drug metabolism bioanalytical laboratory. Readers may add to the list.

In the field of drug metabolism, samples tend to be rather small. Thus, serial bleedings from the tail of a rat yield only 100 μl for each sample. Where the sample size is not small, e.g. urine, the concentration of the component to be measured is often very low and there are usually a formidable number of endogenous compounds. Clearly the sensitivity of an analytical method depends upon the signal-to-noise ratio, but it is not until the signal is fully amplified that the diverse nature of the noise is revealed.

Many of the problems we have encountered in our own laboratory are well known to others who work with biological samples, and have appeared in the literature over the years in a somewhat scattered manner. Reid [1] and Scales [2] reviewed some of these points which arose at the 3rd Bioanalytical Forum in 1979. However, there is much ignorance of such problems amongst colleagues who carry out HPLC on similar samples at high analyte concentration in various chemical and pharmaceutical development laboratories where the knob on the detector amplifier is rarely turned past the halfway mark.

PROBLEMS ASSOCIATED WITH SAMPLE PREPARATION

Sample loss

In dilute solution losses of a given magnitude will be more important, as they represent a greater proportion of the analyte and, in addition, some unwanted effects may proceed faster than in concentrated solution. Thus it is necessary to protect such samples from the influences of light, dissolved gases, heat and adsorption upon glass or plastic surfaces. A slight change of pH can occur at low concentrations with disastrous results. We have found that some compounds which seem reasonably stable in acid solution can, at a low concentration in dilute acid, be unstable if the solution is stored in a soda glass container, which points to release of alkali by the soda glass.

Contamination

Problems of contamination, especially by plasticizers, are well known, and are particularly troublesome in fluorescence work. We are probably all aware of the dangers to our assay results from the optical brighteners in the paper tissues with which we wipe our glassware. Fibres from our "whiter than white" laboratory coats are a constant menace. We all avoid solvents from plastic containers or from containers that have plastic liners to the screw caps, and we hope that the manufacturers did not fill the bottles through a plastic funnel. Problems can also arise through the use of plastic tubing to convey nitrogen gas for evaporating extracts, as found fully 3 years ago in our own laboratory with both HPLC (Fig. 1a) and with GC (Fig. 1b): the source was identified as the PVC tubing conveying nitrogen from the line.

Evidently a tubing component, eventually identified by GC-MS as tributyl aconitate, was volatile enough to pass down with the nitrogen gas but then remained with the sample in the dried-down tube. We were able subsequently to avoid the problem by replacing the PVC with rubber tubing. Hooper & Smith [3] have found a similar problem using PVC where a phthalate was the offending contaminant, but Feyerabend & Russell [4] on the contrary had to replace rubber by plastic tubing to avoid an unidentified contaminant from the nitrogen line.

Glassware cleanliness

One very difficult problem is adequate cleaning of automatic-sampler microvials (e.g. for HPLC 'WISP' use). We have found it vital to include an ultrasonic cleaning step in our washing procedure for such items.

PROBLEMS ASSOCIATED WITH THE HPLC ASSAY

Sample carry-over

Because our assays call for injection of almost the whole sample without using some for pre-rinsing the HPLC sample-loop, we have hitherto been restricted to the use of two types of automatic sampler

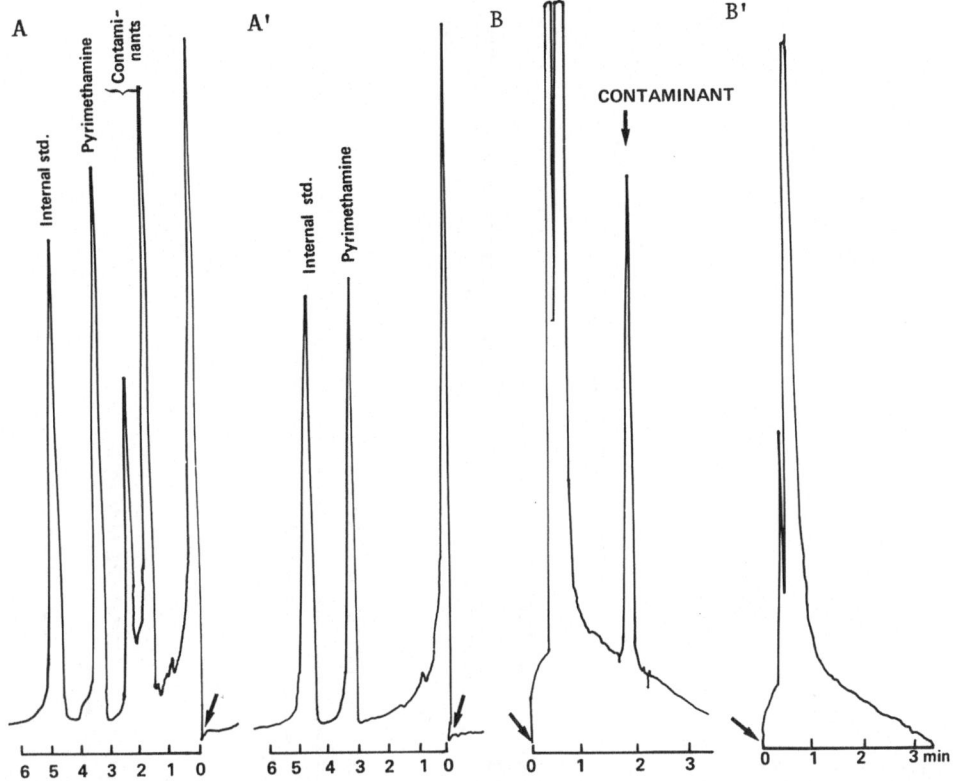

Fig. 1.   A & A': HPLC trace of blank plasma extracts with detection by fluorescence.   **B & B':** GC–ECD trace of plasma extracts containing pyrimethamine.   **A** and **B:** extract dried down under nitrogen conveyed by PVC tubing.   **A'** and **B':** extract dried down by heating, in a water bath (**A'**) or on an oil bath (**B'**). *Arrow* denotes injection.

for HPLC, the Hewlett Packard and the Waters WISP. Both have their own merits and demerits, but the WISP does have the advantage that sample carry-over is immeasurably low. Using the Hewlett Packard it is necessary to insert a wash sample between a high and a low sample since some carry-over occurs on the outside of the needle (Fig. 2).

    It is also instructive to fill an HPLC syringe with a highly coloured substance such as fluorescein solution and then to try washing it out by repeated filling and emptying.

Contaminants from equipment

    Three examples come to mind. A new pressure gauge of the Bourden tube type released fluorescent material for weeks despite pumping litres of hot solvent through. A spot of oil added thoughtlessly to a squeaking metal piston on a reciprocating pump gave

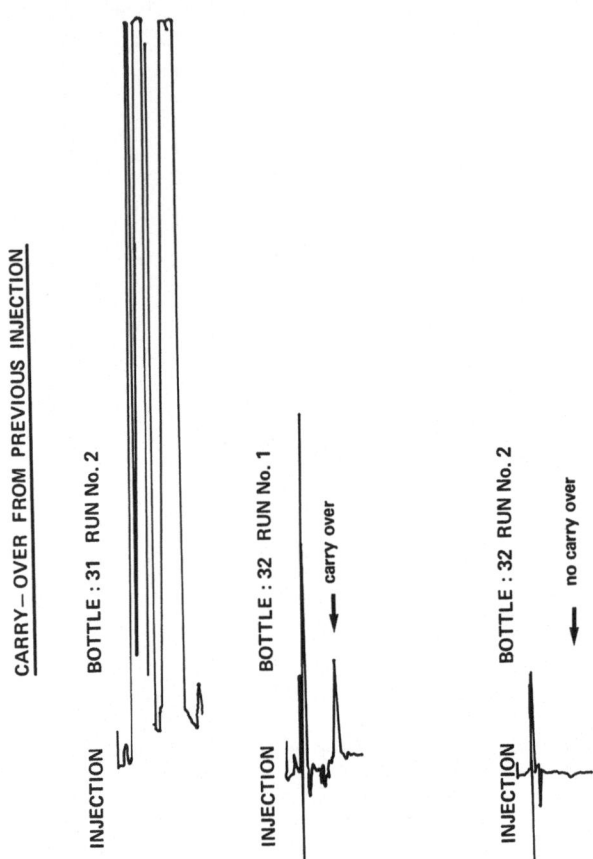

Fig. 2. Carry-over from the previous injection.

similar problems. With a new WISP we found that at each injection a UV-absorbing substance was released through the column. The source was not traced, but was believed to be the seals on one of the valves (Fig. 3).

Naturally our chemist colleagues working with the same substances have not noticed any of these problems because they work at much much higher concentrations.

Solvent background

Clearly the signal-to-noise ratio will be maximal when the mobile phase has a low background value for whatever type of detector is to be used, so it is worth going to some trouble to improve the quality of the mobile phase. For example we were using a mobile phase of methanol/water/NH_4OH for a fluorescent assay and found that both the water and the ammonia were fluorescent. We prepared our own pure water and used this to make our own ammonium hydroxide from ammonia gas, thereby improving the signal-to-noise ratio by a factor of 10.

Fig. 3. Contaminant released by WISP sampler. The volume of mobile phase injected was 15 µl.

Solvent gradients

We rarely attempt to use water/methanol gradients under conditions of high sensitivity. If the water is not pure then impurity peaks will be eluted by the methanol. Even if the water is pure, the pressure changes caused by the viscosity changes will play havoc with the baseline (Fig. 4).

Diluent peaks

The nature of the diluent used for the sample can be very important. Negative or positive peaks can arise. We try to dissolve our samples in the same batch of mobile phase as in the assay, but at high sensitivity even this can give peaks, especially if the mobile phase is being recycled.

Electrical mains spikes

We didn't realise how common these were until we had our fluorimeter set to high sensitivity. A spike occurring on any part of a peak will cause the integrator to interpret this as another peak. Mostly these peaks were of unknown origin but some offending pieces of apparatus were banished from the laboratory because of the havoc they caused. Examples are a 'Whirlimixer' and an old-fashioned automatic analyzer which incorporated electro-mechanical relays; also a xenon lamp if switched on nearby.

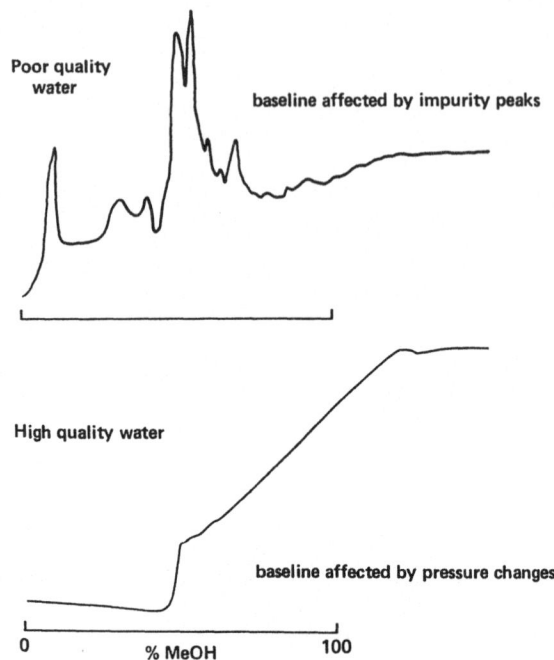

Fig. 4. Runs with a water/methanol linear gradient. UV detector at 254 nm attenuated by a factor of 32.

Endogenous compounds

If any baseline is put under the microscope, as it were, it will be found to be far from flat. This is especially true for samples of biological origin.

Large injection volumes

We regularly use 200 µl when we need to inject an entire sample. To save solvent and also to maintain good equilibrium we also tend to recycle our mobile phase. If the diluent in the sample is different from the mobile phase the composition change can result in considerable changes in the chromatography in the course of time.

Baseline drift

We carry out many of our analytical runs overnight, leaving the automatic system to cope. If the system does not have an autozero incorporated, the baseline will drift off the top or the bottom during the night and we see no chromatograms. The autozero device however is only of cosmetic value as it merely resets the chart recorder to zero at the start of each injection. What is needed is for the *detector* to be reset to zero at each injection. For example our fluorimeter is only capable of giving an output of 10 mV however much light is falling upon the photocell. At the end of an overnight run the baseline output may have risen to 9.5 mV: thus a complete

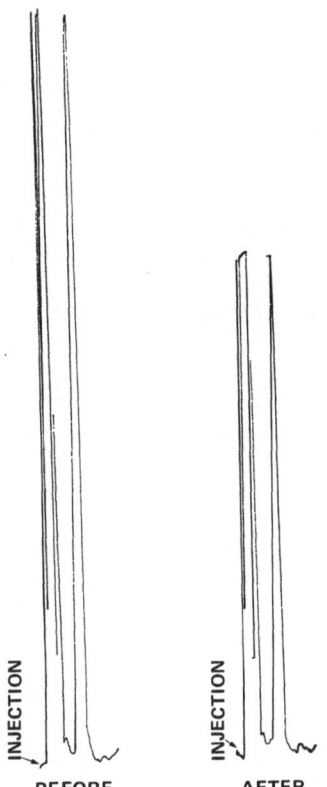

Fig. 5. Effect of baseline drift in an HPLC assembly which did have a recorder autozero.

full-scale deflection is not possible, as only 0.5 mV is available to the pen. The result is that during the course of the run the peaks get smaller and the response becomes completely non-linear (Fig. 5).

Band width

One of the ways of increasing the sensitivity of a fluorimeter is to increase the slit widths and enable more light to pass. It should be remembered, however, that this also increases the spectral band width, which in turn means that the selectivity of the method may be impaired.

Settling down the system

For work of the highest sensitivity we tend to set the system up in the morning with the mobile phase being recycled, and just leave it going. The baseline noise will often decrease dramatically over a few hours, probably due to the detector warming up (Fig. 6). The condition of the detector is an important factor, and

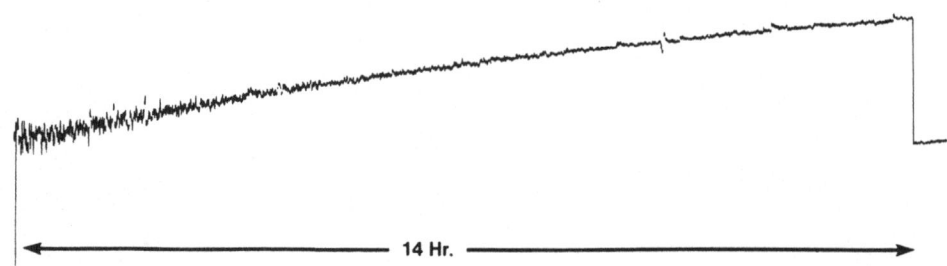

Fig. 6. Fluorimeter baseline pattern during settling-down period.

for very exacting work it may be necessary to replace an old lamp to
reduce the noise.

Pump noise

   Pump noise is usually not apparent till the detector is set to
maximum. Problems are often due to a faulty or a dirty check valve.

Reservoir mixing

   We sometimes find that the baseline is steadier if the mobile
phase in the reservoir is kept stirred. This may be due to evaporation of volatile components or to the arrival of other material
during recycling.

Integration

   This is the commonest cause of problems with low-concentration
samples. Whatever the criterion used by the integrator to distinguish the peak from the background, it will be more difficult at low
signal-to-noise ratios. Commonly the peak is not seen at all, or
the integration starts too late and finishes too early, or the peak
is chopped into several parts by the noise. When peaks are small,
we favour using a ruler to measure the heights.

References

1. Reid, E. (1981) in *Trace-Organic Sample Handling* (Vol.10, this
   series; Reid, E., ed.), Horwood, Chichester, pp. 15-31.
2. Scales, B. (1981) *as for* 1., pp. 353-363.
3. Hooper, W.D. & Smith, M.T. (1981) *J. Pharm. Sci. 70,* 346-347.
4. Feyerabend, C. & Russell, M.A.H. (1980) *J. Pharm. Pharmacol. 32,*
   178-181.

#A-4

EXTRACTION OF DRUGS AND METABOLITES FROM PLASMA AND URINE

M.J. Stewart

Department of Pathological Biochemistry
Royal Infirmary, Glasgow G4 0SF, U.K.

*With the advent of chromatographic separation methods for drug assay, and of notably specific and sensitive detection methods such as RIA, the development of new extraction procedures for determining drugs and metabolites in biological fluids has been relatively neglected. Yet there have been advances in extraction procedures, and this article summarizes the options available and gives examples of their use. Deproteinization, solvent extraction and the use of resins are considered.**

The traditional approach [1] has consisted of an extraction step followed by a detection procedure which varied according to the nature of the molecule. Advances since the 1950s, when chromatographic techniques became available and allowed similar molecules to be separated after extraction and before measuring the 'signal', have led to a range of assay methods that allow drug assays on plasma and urine without extraction or separation steps. These methods, mainly immunological, suit for high throughput and some, e.g. enzyme immunoassay and fluoroimmunoassay, are so rapid as to have been adopted for selected applications. Meanwhile, little effort was put into improving extraction methods, and reviews appeared only rarely [e.g. 2] until the late 1970s. In many situations an initial extraction procedure is advisable or mandatory.-
1. Immunoassay not available or non-specific.
2. Concentrate required to give adequate sensitivity.
3. Endogenous compounds co-react in the final analytical step.
4. Metabolites co-react in the final analytical step.
5. Final analysis by GC or GC-MS requires the analyte to be in an organic solvent.
As is now considered, deproteinization or true extraction is needed.

---

* For the benefit of uninitiated readers this survey has been little abridged; complementary surveys are to be found in Vols. 5, 7 & 10 of this series (and review citations on p. 15 of Vol. 10) in which the most helpful index entries are 'Adsorbents', 'Deproteinization', 'Sample'.–*Ed.*

DEPROTEINIZATION

This procedure may itself suffice if the concentration of the drug in the deproteinized solution is high enough to provide a good signal-to-noise ratio in the final detection procedure. There must be no small water-soluble endogenous compounds which interact in the final reaction, nor any alteration or destruction of the drug during deproteinization. Moreover, the final solution must be appropriate for the final step.

Studies by Jarvie, Fell and myself [3] exemplify the use of de-proteinization alone in measuring a compound in the µg/l range, viz. paraquat – an unusual and highly polar molecule which is converted into a reduced (blue) form by sodium dithionite in an alkaline medium. For clinical reasons it had to be measured in plasma at ~10 µg/l. It can be quantitated at this concentration in aqueous solutions by using derivative spectrophotometry to enhance the signal generated by the molecule. With an adequate signal, the problem resolves into one of producing a water-clear solution in which the coloured ion is stable, without a large dilution of the sample. We found markedly different results with different protein precipitants, amongst which sulphosalicylic acid proved particularly satisfactory (Table 1).

Simple deproteinization techniques, enhanced by the use of high speed centrifugation or millipore filtration to remove the precipitated protein, are especially advantageous in rapid sample preparation for RP-HPLC. An example of such a rapid method is that of Adriaenssens et al. [4] for paracetamol. It was miniaturized [5] to give a total analysis time of only 10 min.

As major drawbacks of protein precipitation procedures, where the total amount of drug in plasma is to be estimated, protein-bound drug may be incompletely released, or released drug may be mechanically trapped in the precipitate.

Removal of protein and measurement of drugs in the supernatant without adding a precipitant is becoming increasingly popular using specially designed micro-filtration techniques. Here the aim is to estimate the concentration of free drug in the plasma. Many of the problems associated with earlier methods, e.g. gradual release of bound drug from the protein during the slow centrifugation or failure to appreciate the pH changes which occur when the protein anions are removed from plasma, have now been overcome or minimized, and reproducible methods are available. Filter systems are marketed by SYVA (EMIT TM free level filter), Amicon (MPS-1) and Worthington (ULTRA-FREE).

SOLVENT EXTRACTION PROCEDURES

In situations such as the five listed above, mere deproteiniza-

Table 1. Effect of precipitant on colorimetric estimation of paraquat (PQ) in plasma.

Protein precipitant	Relative vol. of supernatant solution	Background absorbance (396 nm)	Recovery of added paraquat dichloride
Ethanol	1.4	0.149	43%
Trichloroacetic ac.	2.7	0.067	TCA reacts with PQ^+ radical
Perchloric ac.	3.2	0.123	ClO_4^- reacts with PQ^+ radical
Sulphosalicylic ac.	3.2	0.042	90%
Tungstic ac.	2.3	0.024	Gives blue colour with dithionite in absence of PQ
Zinc hydroxide	1.9	0.013	59%

tion is unsuitable. Solvent extraction can itself entail problems.-
1. Emulsion formation.
2. Interaction of solvent or impurities with the analyte.
3. Introduction of impurities from the solvent, e.g. phthalates.
4. Degradation or loss of analyte during solvent evaporation.
5. Time-consuming solvent evaporation step.
6. Increasing costs of pure solvents.
7. Need for safety precautions when evaporating solvents.

A major advantage of solvent extraction is often stated to be the selective removal of a drug from its more polar metabolites, and misleading clinical information was produced in many laboratories when non-appreciation of the need to selectively extract paracetamol in the presence of polar metabolites led to plasma concentrations being overestimated by up to 400% [6]. However, when urine samples are extracted some metabolites may be extracted into even mildly polar solvents. Volatility problems with administered sodium valproate are minimized by extracting it as its potassium salt. Chlormethiazole is extracted along with volatile metabolites, and the final estimation may fail to differentiate between these; its levels found by the standard UV and GC methods are falsified by co-determination of a readily extracted metabolite [7].

Even if the above problems do not occur with the traditional use of solvent extraction for a non-polar drug, a large volume may be needed for efficient extraction, whereas small volumes are preferable to avoid losses due to adsorption onto glass surfaces, introduction of impurities, and the effects of heat during evaporation of the solvent.

With a suitable internal standard (i.s.), incomplete extraction is allowable. Lest extractability be very different with the chosen solvent [8], this must be thoroughly checked; where no suitable i.s. if available off the shelf, a synthesis is required. An i.s. is especially useful in GC-FID. With a selective, more critical GC detector such as ECD or AFID (NPD), where drug and i.s. may differ in sensitivity, an i.s. for extraction correction, may lead to a false sense of security. Mere replacement of H by D atoms as in GC-MS is an ideal situation.

An example of a short cut where, with i.s. use, incomplete recovery is acceptable is the microextraction pioneered by the Poisons Unit at New Cross [9] with 250 µl Dreyer tubes: rapid equilibration between plasma and solvent is followed by microcentrifugation at high g which removes most emulsions and provides a bead of solvent for GC injection. For highly lipid-soluble drugs such as amphetamines or chlormethiazole a concentration step may be introduced by 'buzzing' 5 ml of urine with 500 µl of solvent. Plasma or urine is commonly subjected to pH adjustment or salt addition so as to increase drug extractability. Morphine and its allied metabolites are a notable exception since their amphoteric nature determines the extraction pH.

Ion-pair extraction

The ion-pairing approach [10], as increasingly used in RP-HPLC and more recently in NP-HPLC, is only now catching on for extraction purposes. Thus, for paraquat we showed in 1978 [11] that this highly polar cation extracts completely into non-polar solvents when paired with sodium dodecyl sulphate. A wide range of cations and anions of graded polarity is available to tailor the extraction of an ionized species which may be recalcitrant to normal solvent extraction. It is perhaps surprising that this approach has not yet been applied to morphine in its cationic or anionic form.

With stable ion pairs there may be a gain in UV absorption, fluorescence or reactivity with a colour reagent, as demonstrated [12] for a wide variety of drugs. Parallel use of an i.s., if carefully chosen, is not precluded. Recovery of the free drug may be effected by a counter-ion. Disadvantages of extracting as an ion pair relate to the difference in signal from that generated by the free drug. The measured analyte must be in one form or the other and not a variable mixture of paired and unpaired drug. The choice of pairing ion also depends on whether the drug has to be recovered from the organic phase in an unpaired form. Solvent extractions may be automated using continuous flow systems such as the Technicon FAST-LC. Such extractions may show high precision due to the reproducibility of treatment of each sample [13], but are time-consuming to optimize and are of value only where sample number for the analyte is large.

LIQUID–SOLID EXTRACTIONS

These feature in other articles in this and previous volumes, as an increasingly practised way to overcome solvent-extraction problems. There are old observations with (e.g.) sodium sulphate or siliconized paper as used to adsorb water from organic phases before evaporating to dryness: drug adsorption onto the solid surface could occur, especially at low concentration in the solvent. There has long been limited use of ion-exchange resins, as columns or as beads which sediment rapidly without centrifugation; but the advent of non-ionic resins such as XAD-2 that can bind non-polar compounds has been of notable benefit. Non-ionic column extraction offers a number of advantages over solvent extraction.-
1. No losses such as emulsions cause.
2. High extraction efficiency and recoveries.
3. Minimal introduction of contaminants.
4. Economical reagents (re-usable).
5. Reduced safety hazard.

Column extraction techniques are easily miniaturized. A Pasteur pipette makes an easily packed disposable column support. Columns of various makes and sizes are now on the market, capable of adsorbing drugs from neat urine or from plasma or blood that has been minimally diluted. Elution is done with a relatively inexpensive solvent, with a much lower volume than for a conventional extraction. Complete automation, including final drying, is achievable with the DuPont PREP-1 system that was transiently on the U.K. market.

Another approach now available is affinity chromatography using a column-bound antibody, which governs the specificity. Other similar molecules may cross-react. Yet one hopes for notable help for extracting some of the more recalcitrant drugs such as antibiotics, despite possible individual anomalies where the patient's own serum carries antibodies such as may occur for penicillin.

CONCLUDING COMMENTS

For many drug analyses there is now a wide choice for the extraction stage. It must, however, be carefully chosen so as to complement the final detection method chosen. Methods are still published in which an excellent and sensitive detection technique is allied to a somewhat crude extraction step. An understanding of the choice now available should lead to more economical, faster and more accurate and precise analytical methods.

References

1. Tompsett, S.L. (1968) *Analyst 93*, 740-748.

2. Jackson, J.V. (1975) in *Isolation and Identification of Drugs*, Vol. 2 (Clarke, G.C., ed.), Pharmaceutical Press, London, pp. 914–920.
3. Jarvie, D.R., Fell, A.F. & Stewart, M.J. (1981) *Clin. Chim. Acta 117*, 153–165.
4. Adriaenssens, P.I. & Prescott, L.F. (1978) *J. Pharm. Pharmacol. 6*, 87–88.
5. Buchanan, T., Adriaenssens, P.I. & Stewart, M.J. (1979) *Clin. Chim. Acta 99*, 161–167.
6. Stewart, M.J., Adriaenssens, P.I., Jarvie, D.R. & Prescott, L.F. (1979) *Ann. Clin. Biochem. 16*, 89–95.
7. Stewart, M.J. & Street, H.V. (1980) *Clin. Chem. 26*, 985–986.
8. Curry, S.H. & Whelpton, R. (1978) in *Blood Drugs and Other Analytical Challenges* [Vol. 7, this series] (Reid, E., ed.), Horwood, Chichester, pp. 29–42.
9. Ramsey, J. & Campbell, D.B. (1971) *J. Chromatog. 63, 303–308*.
10. Schill, G. (1978) *as for* 8., pp. 195–206.
11. Jarvie, D.R. & Stewart, M.J. (1979) *Clin. Chim. Acta 94*, 241–251.
12. Van Buuren, C., Lawrence, J.F., Brinkman, U.A. Th., Honigberg, I.L. & Frei, R.W. (1980) *Anal. Chem. 52*, 700–704.
13. Van der Wal, S. & Snyder, L.P. (1981) *Clin. Chem. 27*, 1233–1240.

#A-5

THE USE OF LONG CAPILLARY GC COLUMNS FOR URINARY PHENOLIC ACIDS FROM CATECHOLAMINES AND ANALOGUES

B.L. Goodwin

Bernard Baron Memorial Research Laboratories
Queen Charlotte's Hospital
London W6 OXG, U.K.

Glass capillary columns are routinely being coated with a phase mixture based on SE-30 Ultraphase for use in analysis of urinary phenolic acids, using 100 m columns of 0.5 mm bore. These have a resolving power of ~10^5 plates, depending on the compound analyzed. An account is given of the preparation of these columns and of their use in examining urinary constituents (human, rat) and in metabolic work with compounds related to catecholamines. When measuring previously unauthenticated compounds, identities must be confirmed by independent means because some of the numerous urinary constituents may interfere.

With the introduction of capillary columns, GC has become a practicable method for the measurement of aromatic acids, but with the disadvantage that columns have a short life (a few months of continuous use at the best) and the commercially available coated columns are so expensive that most laboratories would regard capillary GC as an expensive luxury. Such columns can now be prepared and re-used with a minimum of expense and labour in any laboratory, needing only a vacuum desiccator, a water-pump and a water-bath and a supply of clean compressed air. Procedures are given below.

Measurement of urinary phenolic acids as their TMS or ethyl ester-TMS derivatives [1] needs, for adequate resolution, 100 m of column (0.5 mm bore), suitably as a 13 cm diameter coil. Such columns are physically rugged and the breakage risk is small. The coating is ~10^{-6} m thick, which permits several μl of solution containing up to 100 μg of derivatives to be injected and offers a resolving power of ~10^5 plates, as is essential if the numerous urinary constituents are to be separated. The phase used is SE-30 Ultraphase/OV-25 (4:1). This mixture permits adequate separation of

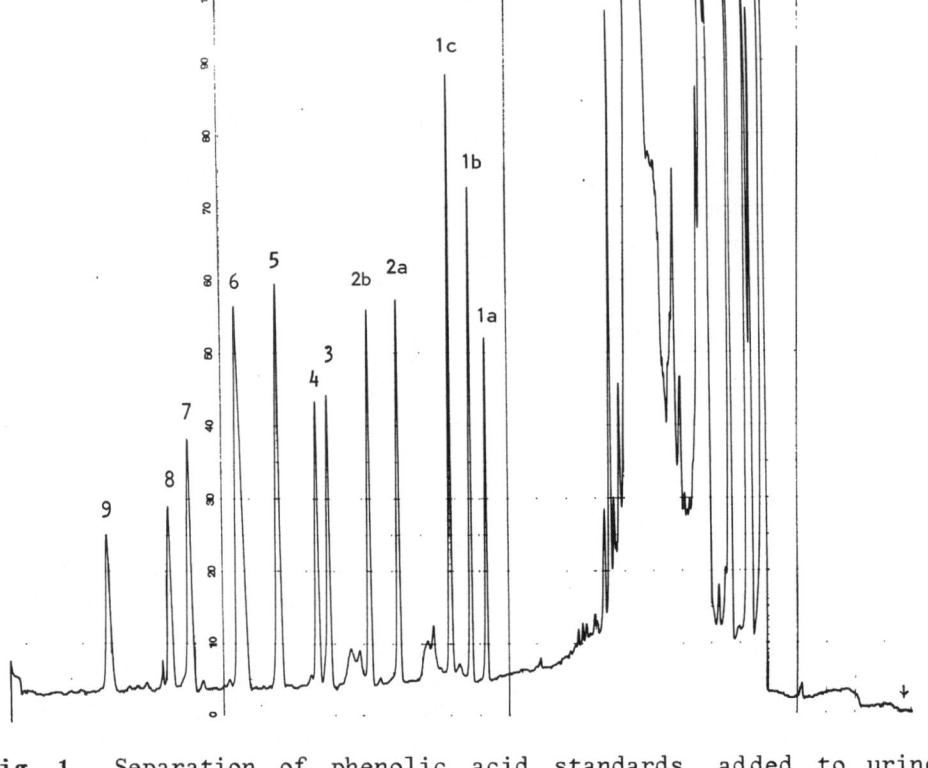

Fig. 1. Separation of phenolic acid standards added to urine, at 210° on a WCOT column 100 m × 0.5 mm i.d. (attenuation 512). Main peaks (each representing ~0.8 μg of the compound):
1: o-, m- and p-hydroxyphenylacetic acids - **a**, **b** and **c** respectively;
2: m- and p-hydroxyphenylpropionic acids - **a** and **b** respectively;
3: 4-hydroxy-3-methoxyphenylacetic acid (homovanillic acid, HVA);
4: p-hydroxymandelic acid;   5: m-hydroxyphenylhydracrylic acid;
6: hippuric acid (endogenous);   7: p-hydroxyphenyl-lactic acid;
8: 4-hydroxy-3-methoxymandelic acid (vanillylmandelic acid, VMA);
9: 3,4-dihydroxymandelic acid (DHMA).
The acids, run as ethyl ester/TMS ether derivatives, were extracted in ethyl ester form by a resin procedure described elsewhere [2].

the major peaks; departure from this composition vitiates the separation of either VMA and p-hydroxyphenyl-lactic acid or HVA and p-hydroxymandelic acid. Fig. 1 (abbreviations given in legend) shows a typical separation of standards. For most work a preliminary separation of the acids as their ethyl esters is carried out on a column of anion-exchanger (AG-1 × 4, acetate form, 100-200 mesh) [2], which permits the isolation of phenolic acids, largely free from aliphatic and non-phenolic aromatics with the exception of hippuric acid (Fig. 2)

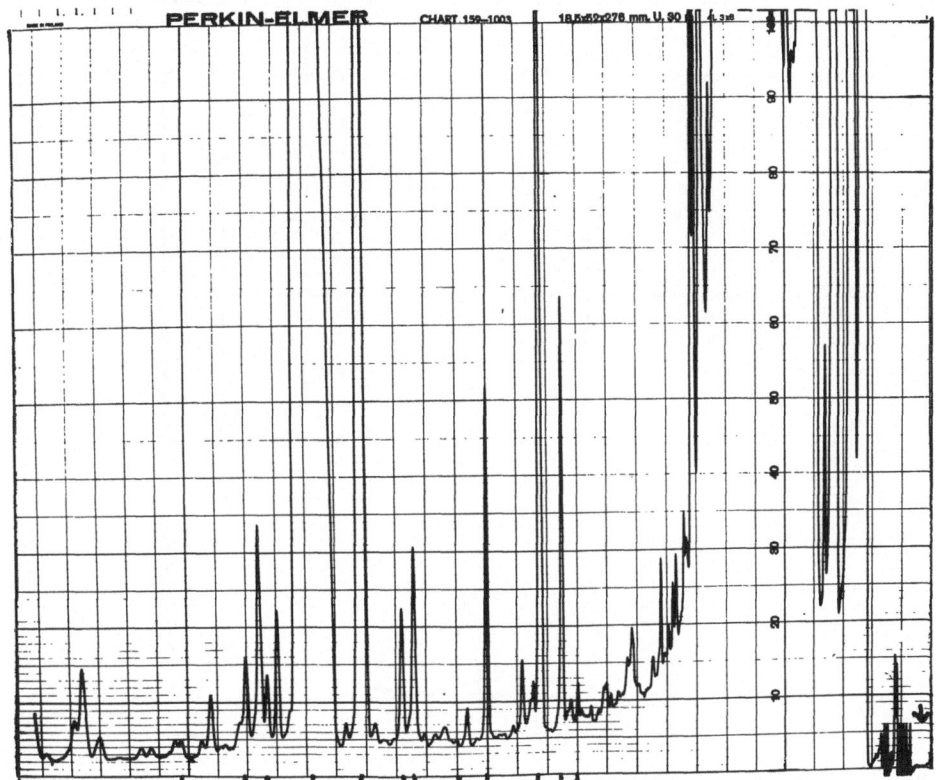

Fig. 2. Extract from human urine, chromatographed as in Fig. 1, demonstrating the presence of the compounds shown in Fig. 1 (their positions being marked below the baseline). Attenuation 128. (For observed levels, see Addendum.)

which, if its concentration is low, tails so much that it does not elute as a peak. The presence of many minor peaks demonstrates unambiguously that the use of packed columns is quite unsuitable except for the measurement of the major urinary components.

The detection system used for the above work was FID. Although capillary chromatography is a great improvement over packed-column GC, it is not the ultimate answer, since with capillaries retention times vary and the chromatograms show so many peaks that their identities may be in doubt. A more specific detection technique is sometimes essential, and this is provided by MS; it involves technical difficulties but can in principle measure as many as 3 overlapping compounds in a peak.

The present approach has been used for examining the profile of organic acids in urine from patients with a range of conditions, e.g phaeochromocytoma, alcoholism, anorexia and hyperactivity. Patterns

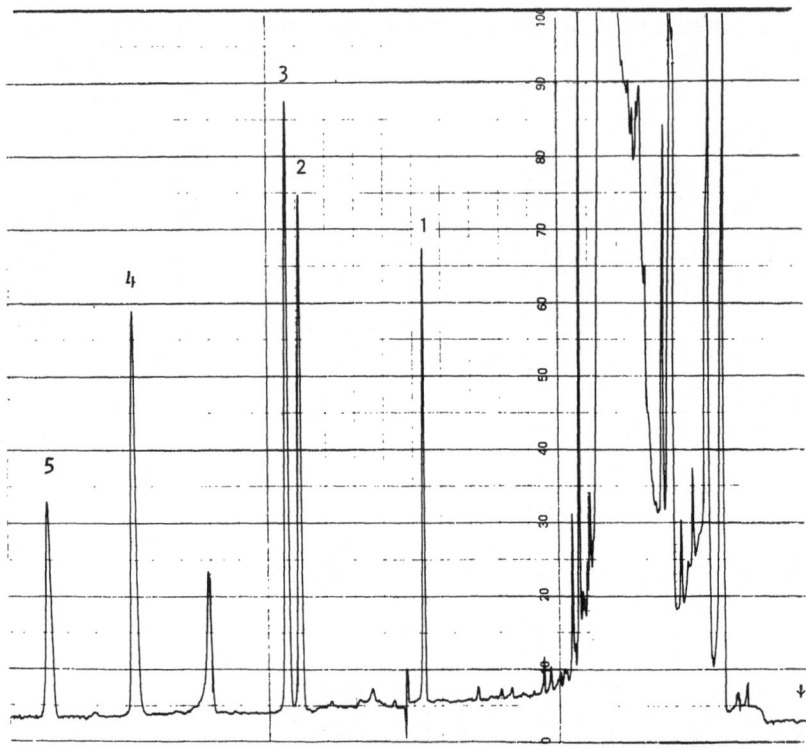

Fig. 3. Alcohol standards added to human urine, as TMS derivatives.
The injection represented 0.6 µg of each compound. Run at 195° on a
60 m column. Attenuation 512 for first peak and then 256.
1: *p*-tyrosol; 2: 4-hydroxy-3-methoxyphenylethanol; 3: *p*-hydroxy-
phenylglycol; 4: 4-hydroxy-3-methoxyphenylglycol (HMPG); 5: 3,4-
dihydroxyphenylglycol.

are qualitatively near-normal, with a very few exceptions; but some
quantitative differences are evident. Thus, the raised VMA and HVA
excretions in phaeochromocytoma are quite apparent. This study has
been extended to metabolic work with rats in which compounds related
chemically to dopa and the catecholamines have been administered and
the excreted metabolites have been measured [3] using, in general,
TMS ether esters for chromatography without separation of aromatic
acids; many of the metabolites would not have been isolated by the
method [2] used in the context of Fig. 2. Except for a few benzoic
acids, the metabolites were mostly measurable. Some acids with a
long side chain (≥6 carbon atoms) were measurable only with a shorter
column (~60 m). It has also been found feasible to assay amines (as
their PFP derivatives) such as β-phenylethylamine and analogues, and
and alcohols such as HMPG, on these columns (as in Figs. 3 & 4). In
Fig. 5 the separation of urinary benzoic and phenylacetic acids is shown.

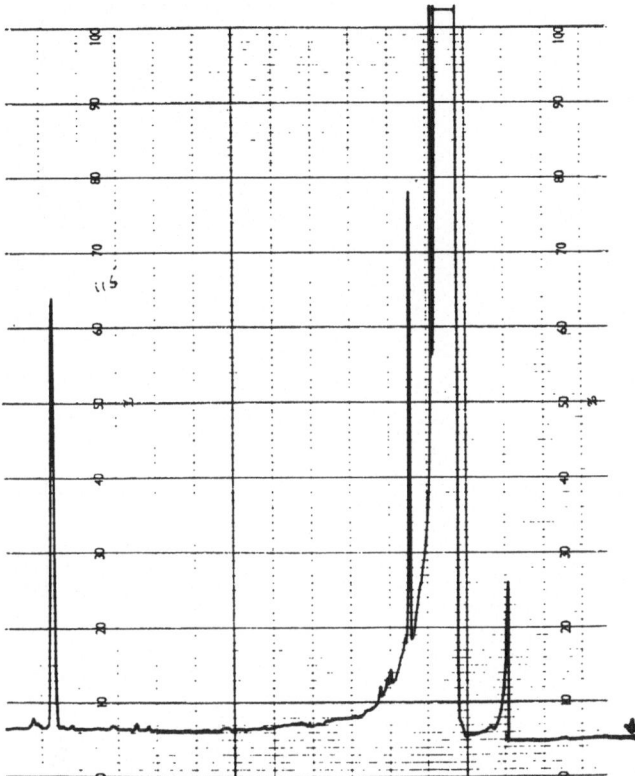

Fig. 4. Authentic *m*–methoxyphenylbutylamine added to rat urine as its PFP derivative. The injection represents 1 µg of compound. Run at 210° on a 60 m column (too hot to distinguish phenylethylamine from solvent peak).

CHROMATOGRAPHIC CONDITIONS

The oven temperature is 210° (but can be varied, e.g. 180° for ethyl phenylacetate), with the injection port and the manifold at 250° using a Perkin–Elmer 900 gas chromatograph with FID. The carrier gas is nitrogen, with a nominal pressure of 0.1 psi/m of column. This permits a gas flow of ~15 cm/s, which is near the minimum compatible with turbulent flow; optimal resolution is obtained near the flow rate at which the flow changes from laminar to turbulent. A make-up gas flow through the manifold is required to optimize the detector response. No special injection port is required since the injection technique permits the complete transfer of the sample to the column prior to elution. Ferrules are made by compressing strips of teflon tape.

Injection is performed by a modified Grob technique that was

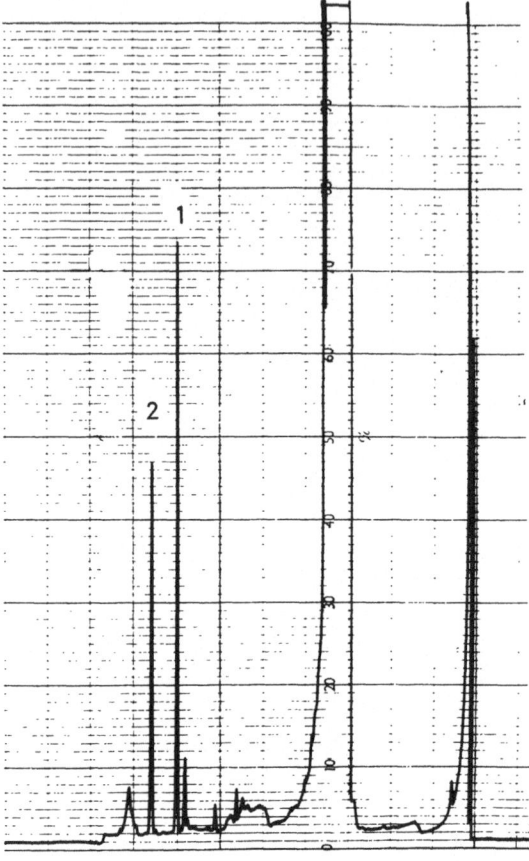

Fig. 5. Benzoic acid *(main peak)* & phenyl-acetic acid *(second peak)* as ethyl esters from human urine after acid hydrolysis. Run at 180° on a 60 m column.

developed independently of Grob [4]. The oven door is opened, and the column is allowed to cool to a temperature (~100°) that is determined experimentally for each type of assay, a temperature at which the compounds being assayed condense on the first few cm of the column, but the fitments at the top of the column have not cooled down sufficiently to allow condensation in them. The column is kept cool for 4 min to allow complete condensation and to carry solvents and other volatile materials towards the end of the column. The temperature is then raised rapidly to the working temperature, and the chromatogram is run isothermally.

COLUMN PREPARATION

The glass columns are subjected to the following procedures [5]. To clean an old column, acetone is forced through it with gradual addition of diethyl ether to the acetone so as to raise the ether concentration by about 10% for every 2 ml used. This removes the GC phase, and the slow increase in ether concentration permits the phase to soften and elute slowly. If the column were washed with pure ether the phase would dissolve as a viscous plug which would probably resist further attempts at removal. The column is finally cleaned by passing 2 M HF through it at 50 psi, which removes the surface layer as a dark brown sludge. Clearly there is a degree of irreversible adsorption of some compounds during the lifetime of a column, and it is postulated that this may be the ultimate cause of loss of column efficiency. This cleaning procedure can be used only with borosilicate glass; soda glass is attacked by HF to yield a fine powder. Even a trace of this is sufficient to create an immovable plug in the column. Hence HF solutions must not be brought into contact with soda glass at all when used for cleaning columns.

The clean column is dried by passing some acetone through it, and the surface is inactivated with bis-trimethylsilylacetamide (~2% in acetone), the excess being washed out with fresh acetone. A new column is inactivated in the same way.

Coating procedure

An 0.4% solution of SE-30 Ultraphase in ether is prepared and doped (as necessary) with other phases; we routinely dope SE 30 with 25% of its weight of OV-25.   This solution is evaporated to two-thirds of its original volume *in vacuo* to degas it. It is then warmed, still under vacuum, to 30-35°, and transferred into a suitable container (one that restricts evaporation) maintained at 35-37°; one end of the column is dipped into the solution. A partial vacuum is applied at the other end; the pressure at this end must not be appreciably lower than the vapour pressure of ether at room temperature. A 100 m column takes ~4 h to fill. The end of the column is closed by drawing in ~20 cm of a melted vaseline-histological wax mixture (6:1), with care not to trap any bubbles; a trapped bubble will inevitably cause failure at the next stage.

The column is placed in a vacuum desiccator and allowed to equilibrate, preferably for several days, to ensure the elimination of incipient bubbles. The pressure is then slowly reduced at a rate that ensures negligible temperature changes, suitably 10 mm/min. This operation should be carried out in a room at near 20°. The pressure is reduced to below the vapour pressure of ether and maintained at that level for 24 h. The vacuum pump is switched off when the meniscus first starts to move along the column at ~1 cm/min. After 24 h, the pressure can be reduced further; re-evacuation at intervals is of course essential to remove ether vapour. The resulting column has a coating nominally 1 μm thick.

References

1. Goodwin, B.L., Ruthven, C.R.J. & Sandler, M. (1974) *Clin. Chim. Acta 55*, 111-112.

2. Goodwin, B.L., Ruthven, C.R.J., Fellows, L.E. & Sandler, M.(1976)
    Clin. Chim. Acta 73, 191–197.
3. Goodwin, B.L., Ruthven, C.R.J., King, G.S., Sandler, M. & Leask,
    B.G.S. (1978) *Xenobiotica 8*, 629–651.
4. Goodwin, B.L. & Sandler, M. (1975) *Clin. Chim. Acta 59*, 253–254.
5. Goodwin, B.L. (1979) *J. Chromatog. 172*, 31–36.

ADDENDUM: Mean excretions of phenolic acids in man

The following values (mean ±S.D.) have been found for those compounds
in the legend to Fig. 1 that can be satisfactorily assessed (cf. Fig. 2).

No. in Fig. 1 legend	mg/24 h	mg/g creatinine
1a	1.16 ± 0.44	0.92 ± 0.34
1b	9.5 ± 5.6	8.55 ± 6.9
1c	31.8 ± 16.4	29.3 ± 28.6
2a	2.7 ± 1.5	2.45 ± 2.3
2b	0.45 ± 0.33	0.36 ± 0.24
3	6.35 ± 1.95	5.15 ± 1.8
4	3.75 ± 1.25	2.45 ± 1.0
7	2.0 ± 0.95	1.75 ± 1.3
8	4.65 ± 1.5	3.75 ± 1.3

#A-6

QUANTITATIVE HPTLC IN ASSAYING BODY FLUIDS
FOR DRUGS AND METABOLITES

W. Ritter

Bayer AG, Department of Pharmacokinetics
D-56 Wuppertal 1, West Germany

*High-performance thin-layer chromatography (HPTLC) is an instrumentalized approach, more reliable than conventional TLC, and superior in other respects such as sensitivity. Commercially available instrumentation, e.g. for plate development in its different modes, is considered, and the optimization of a spotting device (see **Appendix**). Examples of applications are given later (art. #D-5).*

HPTLC, entailing use of plates with silica gel of a narrow particle size distribution, is an entirely new analytical system of higher reliability than conventional TLC [1, 2]. For assaying drugs and their metabolites in body fluids, HPTLC has obvious advantages over GC and HPLC:
- up to 40 samples can be assayed simultaneously;
- sensitivity (detection limits) generally much better than in GC and HPLC;
- due to the 'off-line' measurement by scanning, several different assays can be performed simultaneously in the laboratory, using *one* densitometer for quantitation in succession (economy!);
- no restrictions, such as apply to HPLC, in the choice of organic solvents for the mobile phase;
- clean-up steps are less critical than in GC and HPLC, because the TLC plate, in contrast with columns, is used only once;
- post-chromatographic derivatization (much better than pre-chromato-graphic derivatization) to increase sensitivity, selectivity and specificity is much easier to perform than in HPLC, and all samples can be derivatized simultaneously;
- substances being quantitated need migrate only far enough to be separated from interfering materials (whereas in GC and HPLC all the components injected have to be eluted from the column);
- the (HP)TLC plate serves as a document for further investigation, i.e. specific colour reactions after measurement in the UV, or measu-

rement of the total UV spectrum, after scanning at the wavelength of
maximum absorbance, to characterize metabolites;
- spiked samples of the body fluid, as standards, can be spotted and
assayed simultaneously, thus increasing the accuracy and reliability
of quantitation.

    Once the body-fluid samples have been extracted, the instrumen-
talized procedure which follows entails three consecutive steps, viz.
sample application onto the plate, development of the chromatogram,
and densitometric measurement. [#E-4 in Vol. 5, by Faber, is relevant-*Ed.*].

SURVEY OF HPTLC INSTRUMENTATION

Sample application

    Three CAMAG instruments permit sample application as spots: the
Nano-Applicator for spotting of volumes between 50 nl and 230 nl,
the Micro-Applicator for 0.5-2.3 µl, and the Nanomat, which can be
used in connection with fixed-volume capillary pipettes (100 nl or
200 nl) or with disposable micro-capillaries (0.5 to 5 µl) or with
either of the above-mentioned applicators which consist of micro-
meter-controlled syringes. These various instruments are perfect
spotting devices for quality control, but their application for body
fluid assay of drugs and metabolites is limited and only possible
in cases where the concentrations are high and the assay is very
sensitive [3].

    For sample application of larger volumes as streaks or bands,
the CAMAG Linomat III and the DESAGA Autoliner 75 are available.

    Spotting is still a crucial step in quantitative HPTLC, especi-
ally when relatively large volumes have to be applied, e.g. for
pharmacokinetic studies of drugs administered in doses of only a few
mg per person. Under these circumstances, concentrations of drugs
and metabolites in body fluids are very often in the low ng/ml range
and, even with very sensitive assays, volumes of up to 150 µl had to
be applied as spots using microlitre syringes [4].

    An interesting and very different application procedure has
been described, 'contact spotting' [5]. However, as this Contact
Spotter was not - and still is not - available in Europe and we
urgently needed a spotting device to simultaneously apply 15-20 samples in volumes of up to 50 µl as spots, the DESAGA Autospotter was
used. It was not designed as a spotter for high-performance TLC,
nor was this capability claimed by the manufacturer. Nevertheless
we found a way to optimize the Autospotter so as to achieve low
inter-channel differences and spot diameters of less than 2 mm. The
Autospotter's advantageous feature of spotting 20 samples *simultaneously* resulted in a quite dramatic increase in the number of samples
assayed per unit time, obviated the imprecision of manual spotting,

and enabled us to perform pharmacokinetic studies which would never have been possible without this automatic spotting device (see the 'Analytical case histories' late in this book, art.#D-5). *N.B.-See Appendix* on the Autospotter optimization, p. 60.

The CAMAG Five-solvent U-chamber offers the possibility of spotting volumes up to 1000 μl ('enrichment on the plate'); but development is restricted to the circular technique, and only 4 samples are spotted simultaneously.

Three further spotting instruments that can be used for HPTLC determination of drugs and metabolites warrant mention*:
- the AIS spotter from Analytical Instruments Specialities (Libertyville, Illinois) with provision of 10 or 19 syringes which allows up to 10 or 19 samples to be spotted simultaneously [see article by D.A. Stopher, #NC(A)-1];
- the KONTES sample applicator for simultaneously spotting up to 6 samples [6];
- the DESAGA Autodoser for simultaneous spotting of up to 10 samples automatically (or, as streaks or bands, volumes up to 1000 μl).

Another instrument, the BAYER Quantomat for simultaneous high-performance spotting of up to 10 samples (also as streaks or bands) is produced only upon request [7].

Development of chromatograms

The simplest device for chromatogram development, the *conventional* tank, can also be used for HPTLC.   The CAMAG *twin trough* chambers have been proven to be more advantageous, and are used widely in our experimental work.   Another developing device for linear chromatography, a sandwich chamber for *horizontal* development constructed by CAMAG, allows the development of twice as many spots as in vertical development – the development being carried out from both edges of the plate – but separation distance is reduced to less than 5 cm.

For *circular* HPTLC, the CAMAG Anticircular U-chamber is available. The samples – as many as 48 – are spotted on an outer circle, to be developed towards the centre.

Densitometric measurement

The following list is by no means complete, and is confined to instruments which are known to the present author from his own bench experience: it is in alphabetical order of manufacturers, and deals only with reflectance and fluorescence measurement:
- the CAMAG TLC/HPTLC Scanner, available with monochromator or with standard filters, for automatic scanning of linear, circular and

* The 'Wolfson' instrument (Surrey University; p. 62) wasn't surveyed.–*Ed.*

anticircular chromatograms;
- the SCHOEFFEL Densitometer SD 3000 for scanning of linear chromatograms only, with manual positioning of the spots under the light
beam (double-beam instrument);
- the SHIMADZU Dual-Wavelength TLC Scanner CS-910 using the zig-zag
scanning technique, with accessories for automatic background correction and for linearizing calibration curves, and with the opportunity
to change from a dual-wavelength single-beam mode to a single-wave-
length double-beam mode;[*]
- the SHIMADZU High-Speed TLC Scanner CS-920 for automatic scanning
and calculation of the amounts per spot or concentrations per ml
(depending on the standards), the calculation being based on equa-
tions to obtain calibration lines with internal or external standards by a one-point or two-point method: the scanning results for
up to 4 standard spots are averaged to minimize spotting errors, and
and there is additional equipment for fluorescence measurement and
for scanning circular chromatograms;
- the ZEISS Chromatogram Spectrophotometer KM3 (no longer available
because production was discontinued), a true high-performance single-
beam instrument with monochromator.

    In our own experimental studies on the pharmacokinetics and
metabolism of drugs, we prefer linear chromatography, spotting with
the Autospotter, and quantitation is based on scanning with the ZEISS
Chromatogram Spectrophotometer KM3, coupled with an electronic
integrator and a recorder.

POST-CHROMATOGRAPHIC DERIVATIZATION REACTIONS

The main incentive to derivatizing a substance is increase of sensitivity, selectivity and specificity. Quantitative determination by
(HP)TLC of drugs and metabolites in body fluids offers the advantage
that post-chromatographic derivatization is feasible, such derivatization being precluded in GC and not easy to perform in HPLC. Hence
if a substance to be assayed lacks intrinsic fluorescence and if UV
absorbance is insufficiently sensitive or specific - as is common in
body-fluid assay for drugs or metabolites, for various reasons (low
dosage, low bioavailability, extensive first-pass metabolism, rapid
biotransformation, etc.) - then derivatization has to be performed.

    In HPTLC the *post*-chromatographic technique seems preferable
to *pre*-chromatographic derivatization, because all samples on the
plate can be derivatized simultaneously, and no by-products of the
derivatization reaction need to be separated. The simplest application of this technique is often not regarded as such, viz. the treatment of a TLC plate with acid [8] or an oxidizing agent [9] or tri-
ethanolamine [10] to enhance or produce fluorescence. Post-chromatographic derivatization of primary amines with fluorescamine is
another reaction to achieve fluorescence [11,12; cf. #D-5 in Vol. 7].
---
[*] For another 'zig-zag' instrument (Vitatron), see #NC(A)-1.- *Ed.*

Post-chromatographic colour reactions on the TLC plate so as to get stained spots have long been practised. However, although there are numerous colour reactions in handbooks on paper and thin-layer chromatography, only a few have been applied in quantitative TLC, and seemingly none have so far been reported for HPTLC plates (some examples are given as 'case histories' later in the book, #D-5).

The first published example of a post-chromatographic colour reaction for quantitation by ordinary TLC, the so-called 'Iodine-Pauly' reaction for clotrimazole in biological material [13], was carried out with 3 spraying reagents, each spraying being followed by heating the plate in an oven at 100°.

The Bratton-Marshall derivatization reaction [14] was applied for thin-layer densitometric determination of amitriptyline and nor-triptyline in body fluids [15]. The sensitivity, 0.5 ng/ml, was 20-fold better than that of the fluorimetric assay [16]. The Bratton-Marshall reaction was also successfully applied for the assay of metoclopramide, clebopride and several metabolites in body fluids [17]. Aromatic compounds such as chlorpheniramine can be visualized by nitration followed by reduction and coupling with the Bratton-Marshall reagent [19].

For quantitation of paracetamol in plasma, a thin-layer densi-tometric method was reported using the Folin-Ciocalteu reagent for a post-chromatographic colour reaction [19]. A thin-layer densito-metric method based on a colour reaction with Ehrlich's reagent was described for furosemide [20].

The later 'case histories' article brings out the superiority of post-chromatographic derivatization by colour reactions, especi-for sensitivity, as well as the advantages vs. GC and HPLC.

References

1. Zlatkis, A. & Kaiser, R.E. (eds.) (1977) *HPTLC High Performance Thin-Layer Chromatography*, Elsevier, Amsterdam, and Institute of Chromatography, Bad Dürkheim, 240 pp.
2. Bertsch, W., Hara, S., Kaiser, R.E. & Zlatkis, A. (eds.) (1980) *Instrumental TLC*, Hüthig, Heidelberg.
3. Fenimore, D.C., Meyer, C.J. & Davis, C.M. (1977) *J. Chromatog. 142*, 399-409.
4. Ritter, W. (1977) *J. Chromatog. 142*, 431-440.
5. Fenimore, D.C. & Meyer, C.J. (1979) *J. Chromatog. 186*, 555-561.
6. Touchstone, J.C. & Levin, S.L. (1980) *J. Liq. Chromatog. 3*, 1853-1863.
7. Pohl, U. & Schweden, W. (1975) *Z. Anal. Chem. 274*, 265-269.

8. Faber, D.B., de Kok, A. & Brinkman, U.A.Th. (1977) *J. Chromatog.*
    143, 95–103.
9. Søndergaard, I. & Steinness, E. (1979) *J. Chromatog. 162*, 422–426.
10. Schäfer. M., Geissler, H.E. & Mutschler, E. (1977) *J. Chromatog.*
    143, 615–623.
11. Steyn, J.M. (1977) *J. Chromatog. 143*, 210–213.
12. Lieber, E.R. & Taylor, S.L. (1978) *J. Chromatog. 160*, 227–237.
13. Ritter, W., Plempel, M. & Pütter, J. (1974) *Arzneim.-Forsch.*
    (Drug Res.) 24, 521–525.
14. Bratton, A.C. & Marshall, E.K. (1939) *J. Biol. Chem. 128*, 537–550.
15. Haefelfinger, P. (1978) *J. Chromatog. 145*, 445–451.
16. Faber, D.B., Mulder, C. & Man in't Veld, W.A. (1974) *J. Chroma-*
    tog. 100, 55–61.
17. Huizing, G., Beckett, A.H. & Segura, J. (1979) *J. Chromatog. 172*,
    227–237.
18. Haefelfinger, P. (1976) *J. Chromatog. 124*, 351–358.
19. Gupta, R.N., Eng, F. & Keane, P.M. (1977) *J. Chromatog. 143*, 112–114.
20. Steiness, I., Christiansen, J. & Steiness, E. (1979) *J. Chroma-*
    tog. 164, 241–246.

Appendix

OPTIMIZATION OF AN AUTOMATIC SPOTTER FOR HPTLC USE

More widespread use of HPTLC for body-fluid assays in the ng/ml
range [#D-5] hinges on facilitating the slow and tedious spotting
operation. Several high-performance spotters tested in our laboratory have given sufficiently small spot diameters (<2 mm for 10–20 µl
applied) but were no quicker than manual spotting by microlitre
syringe (>1 h for 20 samples). The above-mentioned DESAGA Autospotter, not high-performance, has proved to be near-ideal, if properly
optimized and adjusted, for pharmacokinetic studies where an imprecision of ±10% can be tolerated (comparable to inter- and even intra-
individual differences).

The instrument has 20 channels, viz. 20 horizontal lengths of
tygon tubing each connected to perpendicular teflon tubing which
is the real 'spotter', being the sole locus of the solvent (i.d. ~0.4 mm)
and having the lower 7 cm pulled through a stainless steel needle to
help handling and protect from damage. The tygon elements are simu-
laneously pressed by a pair of steel rollers which move up to 50 µl
in each forward (drawing in) and backwards (expulsion) movement. The
choice of speed, 1–6 µl/min, depends on the solvent's surface tension
and volatility. The plate can be heated from below, or exposed to an
air current, or illuminated for viewing.

OPTIMIZATION

Inter-channel differences should be minimal, as is considered later, and spots must be small, entailing selection of a suitable solvent. The least polar solvent in which the drug is soluble should be selected [9], and interactions between drug and silica gel exploited [10]. Dichloromethane was the best choice for extracting and directly spotting muzolimine, which is taken up immediately by the silica gel to produce a very small analyte spot despite a large solvent area until it evaporates ([4], above). An analyte spot < 2 mm diameter suffices in the present context; we have found a spot of only ~0.1 mm at the origin to attain 2-3 mm even within 1 cm of separation distance. We attributed this to endogenous components from the biological material, but even a pure substance such as Sudan Green dye behaves similarly [11]. In contrast, spots of < 2 mm given by the Autospotter widen to $\not>$2-3 mm with the right developing solvent and with R_f values of ~0.3-0.5.

Moreover, the teflon tips must be an optimum distance from the gel layer, with initial use of an uncoated TLC plate to adjust all tips to the same height. The final optimal distance apart depends on solvent surface tension and volatility, and has to be determined essentially by trial-and-error; the same solvent as is to be used for the drug is used for the test substance – suitably an azo dye, but we prefer the drug itself despite the need for post-chromatographic derivatization. A dye does save time in starting optimization. Once tip adjustment is done, a 20-spot run is performed. Fig. 1A shows typical scanning results: S.D.>10% for peak height or area (integral).

At this incompletely optimized stage the spotter can serve for assays with an i.s. (we seldom practise this) or for semi-quantitative assay as in compliance monitoring of patients, e.g. when muzolimine was being newly studied: daily throughput was 3x(19 tests+1 std.).

Of the 20 spots in quantitative assays on a 20 x 10 cm plate, say 6-10 are standards (and test samples are each put on 2 or 3 plates). The desired inter-channel S.D. depends on the kind of assay. Monitoring dog blood for G-I absorption in a long-term toxicity study may tolerate 8-10% imprecision, whereas drug and metabolite determination needs <±5%, and the S.D. of the inter-channel difference must be ~2%, as in Fig. 1B (where n = 20). If this performance cannot be attained within 2 working days, the most aberrant tygon elements are changed.

Optimization of the Autospotter in respect of inter-channel differences may take up to 3 days, depending on the experience of the operating technician. However, the time investment is repaid insofar as there can be many weeks of use without further attention to optimization. An additional point concerns memory effects when assaying in the low ng range: the DESAGA plastic sample vessels are replaced by glass vessels which can be cleaned more readily.

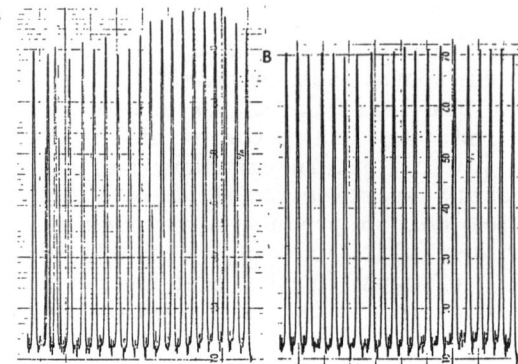

Fig. 1. Scans for 20 spots applied by Autospotter, **A:** not yet fully optimized, and **B:** optimized to perform quantitative (HP)TLC.

BENEFITS OBTAINED

The Autospotter has made possible the following improvements in our HPTLC assay work on body-fluid drugs and metabolites:
- our daily throughput has doubled or trebled;
- assay reliability has increased with the elimination of the imprecision of manual spotting;
- spiked aliquots of the biological material, run simultaneously, have replaced standards consisting merely of drug solutions which, to save time, had been applied in volumes of only a few µl;
- improved assay precision is attained for drugs unstable on silica gel layers, e.g. muzolimine and nafazatrom (#D-5) because standards and unknowns are spotted, and undergo any degradation, in parallel;
- as 20 samples can be done within 20 min (40 µl of extract applied at 2 µl/min), assays can be run in triplicate instead of in duplicate to increase accuracy, precision and reliability.

Altogether the Autospotter has changed our laboratory's TLC work considerably and has become a near-irreplaceable tool in determining drugs and metabolites in body fluids by HPTLC.

Additional references

9. Kirchner,J.G. (1973) *J. Chromatog. 82*, 431-440.
10. DeAngelis, R.L., Robinson, M.M. & Sigel, C.W. (1980) *J. Liq. Chromatog. 3*, 833-839.
11. Fenimore, D.C. (1980) in *Instrumental HPTLC* (Bertsch, W., Hara, S., Kaiser, R.E. & Zlatkis, A., eds.), Hüthig, Heidelberg, pp. 81-95.

Editor s' note on the 'Wolfson' Multi-Sample Applicator.- This instrument [#NC(E)-2 in Vol. 10, this series] can repetitively apply 12 samples simultaneously to give spots of ~2 mm diam. via glass capillaries. It works better with dichloroethane than with dichloromethane (volatile!). Address enquiries to Dr. E. Reid (address in Preface).

#NC(A)

NOTES and COMMENTS relating to

Techniques applicable to metabolite investigation

Comments related to particular contributions:

#NC(A)-1

A Note on

TLC DENSITOMETRY FOR DRUG AND METABOLITE DETERMINATION

D.A. Stopher

Pfizer Central Research
Department of Drug Metabolism
Sandwich, Kent CT13 9NJ, U.K.

Although in our Department we make extensive use of HPLC, GC and GC-MS, we have also found quantitative TLC to be a very useful technique for the analysis of drugs and metabolites in body fluids. This approach is not often discussed but deserves mention since the equipment is simple and the thin-layer plate can accept heavy loading of relatively crude extracts of plasma. Analytical methods can be developed quickly as extensive 'clean-up' is not necessary. The procedure calls for automatic equiment. We use a 'Multi-spotter' (A.I.S.; from Universal Scientific, London) and the Vitatron densitometer which can quantitate compounds on TLC plates by direct absorption in visible or UV light, by fluorescence quenching (a fluorescent substance being incorporated in the absorbent layer), or by fluorescence emission. Examples are now given.

Fluorescence quenching was the means of determining the level, in urine and plasma, of two compounds having sulphone and sulphonamide groups:- UK-12,130, a cerebral vasodilator, and its metabolite UK-13,221. The procedure, which for plasma entailed adding 2 ml of water to a 1 ml sample, comprised two extractions with ethyl acetate and drying down under nitrogen in a tapered tube. The residue was taken up in 100 μl of methanol for TLC.

The calibration curves produced in this assay (Fig. 1) are sometimes non-linear due to deviations from Beer's law. The data from the calibration samples were fitted to a quadratic equation by computer using the method of least squares.

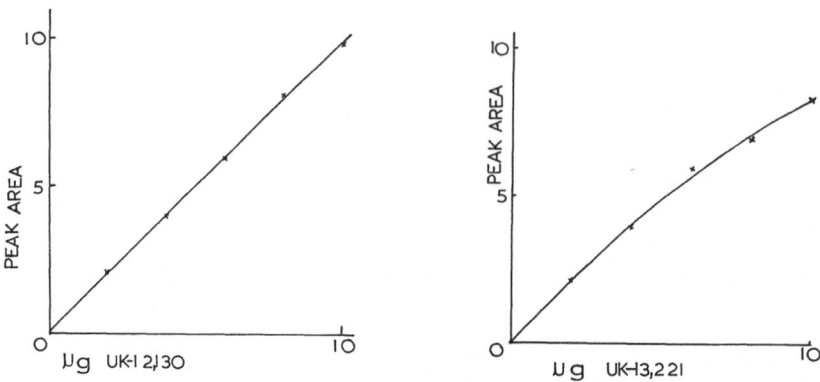

Fig. 1. Calibration curves for UK-12,130 *(left)* and its metabolite UK-13,221 *(right)*. TLC on silica gel with fluorescent indicator; a 72:28:1 mixture of ethyl acetate/2-propanol/0.88 ammonia as eluent.

UK-22,486, the active metabolite of oxfenicine, was measured in urine as a 2,4-dinitrophenylhydrazine derivative by direct absorptiometry, after the work-up procedure shown in Scheme 1. Again the calibration is non-linear (Fig. 2), and a quadratic regression needs to be used to calculate results.

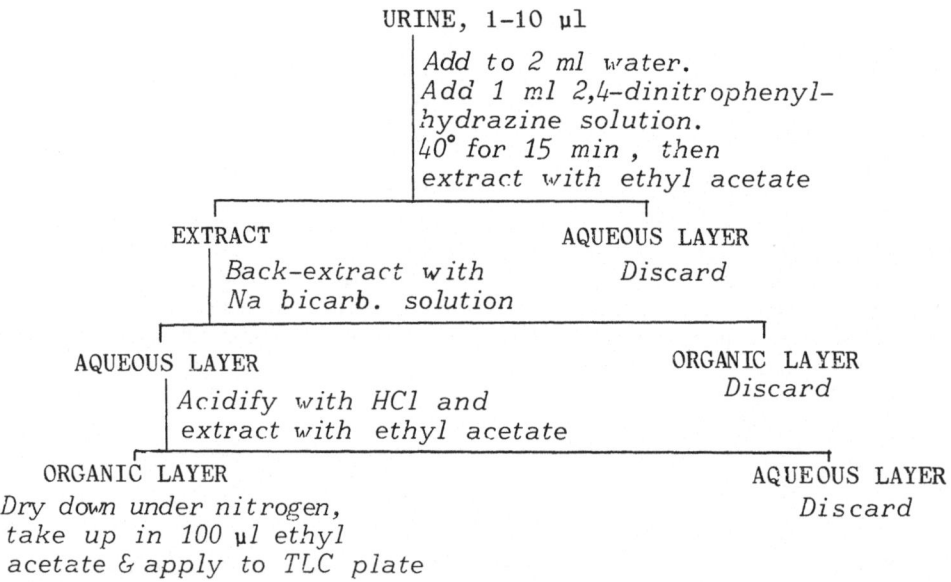

Scheme 1. Work-up procedure for UK-22,486 (an active metabolite).

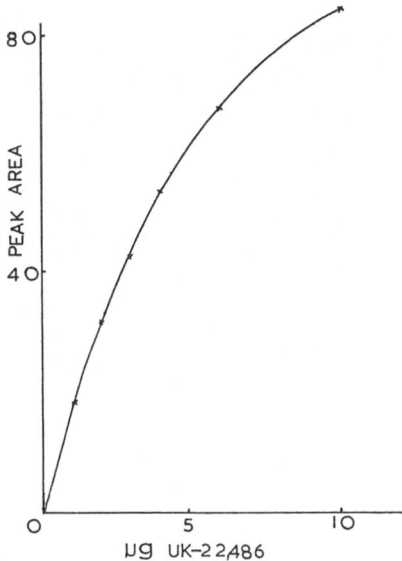

Fig. 2. Calibration curve for UK-22,486.
TLC on silica gel with ethyl acetate/methanol/glacial acetic, 80:20:1.

UK-25,842
OXFENICINE

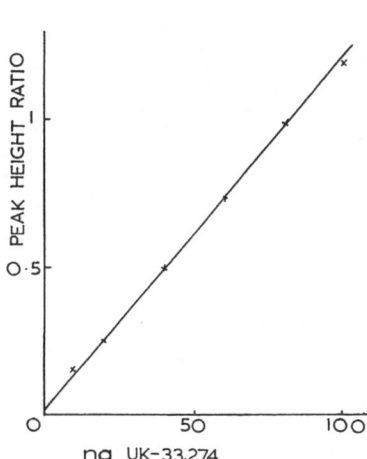

Fig. 3. Calibration curve for UK-33,274.
TLC on silica gel with chloroform/hexane/liquid paraffin / methanol/0.88 ammonia, 50:30: 10:10:0.05.

UK-22,486

For the antihypertensive compound UK-33,274 in human plasma, direct fluorescence was used to provide a sensitive assay. Prazosin served as internal standard, added to 1 ml of plasma followed by 1 ml of KOH solution. A single extraction with diethyl ether, and drying down under nitrogen in a tapered tube, gives a residue which is taken up in 100 μl ethyl acetate for TLC loading. Peak height ratios as measured from the recorder chart furnish a linear calibration curve (Fig. 3). We routinely add 10% (v/v) of liquid paraffin to the solvent system for fluorescence analysis; this remains on the TLC plate and gives a considerable enhancement of the fluorescence. A sensitivity of 10 ng/ml plasma is easily obtained in the assay.

UK-33,274

PRAZOSIN

#NC(A)-2

A Note on

TAKING ALIQUOTS FROM SMALL VOLUMES OF UPPER PHASES

H. de Bree

Duphar Research Laboratories
P.O. Box 2, Weesp, The Netherlands

In the development of a drug assay method we ran into difficulties connected with the use of small volumes. It appeared to be feasible to extract the analyte completely from 2 ml of aqueous solution into 50 µl of isooctane; but to take an aliquot of the organic layer into a syringe for chromatography was troublesome. It was hardly possible to sample the organic phase without contaminating it with the aqueous phase.

As is illustrated (Fig. 1), a simple manipulation using a cone-terminated capillary, made from a Pasteur pipette, enlarges the organic layer thickness from 2 mm to about 20 mm. Thereby even 20 µl may be taken easily from a 50 µl sample extract. The stratagem has also been successfully applied to a chloroform phase above saturated potassium carbonate.

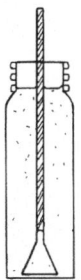

Fig. 1. Use of the device (the chromatography syringe needle is inserted into the capillary).
1: screw-capped vial.
2: cone-terminated capillary.
3, 4: aqueous & organic phases.

#NC(A)-3

A Note on

RAPID REMOVAL OF FINE PARTICLES FROM
ALUMINA FOR COLUMN CHROMATOGRAPHY

B.L. Goodwin,
Bernhard Baron Memorial Research Laboratories
Queen Charlotte's Hospital
London W6 OXG, U.K.

The background to the procedure shown in Plate 1 is that alumina is an essential tool in the measurement of catecholamines and analogues, since they are fairly specifically adsorbed at slightly alkaline pH and are readily eluted at acidic pH. After the requisite acid

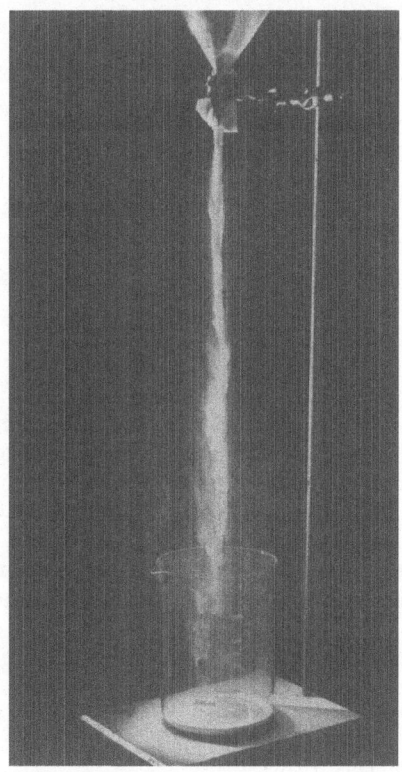

Plate 1. Particle separation, in a draught-free atmosphere. The dry powder is poured through a filter funnel with a circular motion ~1 m above a suitable receiver. It pours through the funnel as an aerosol, and the turbulence created by the flow of powder carries the finer particles outside the rim of the receiver. This process is repeated as often as required to achieve whatever degree of separation is desired, typically when 10% of the total alumina has been lost.

Plate 2. The effect of fractionation on the size ($\vdash\!\!\underline{\quad 400\ \mu m \quad}\!\!\dashv$) of alumina particles. *Above:* prior to fractionation. *On right, above:* after fractionation; *below:* dust recovered.

pre-wash, alumina is a powder with a large range of particle sizes. It is essential to remove the finer particles, since they drastically reduce the flow rate through an alumina column. The traditional way [1] entailing washing with water is inefficient as it removes little more than colloidal material, and is very time-consuming.

The new rapid method as described in the legend to Plate 1 can in principle be applied to any powder. It has been used successfully with the GC support Chromosorb W-HP, which is reduced to a fine powder by any more vigorous technique.

Reference

1. Weil-Malherbe, H. (1961) in *Methods in Medical Research, Vol. 9* (Quastel, J.H., ed.), Year Book Medical Publishers, Chicago, pp. 130–146.

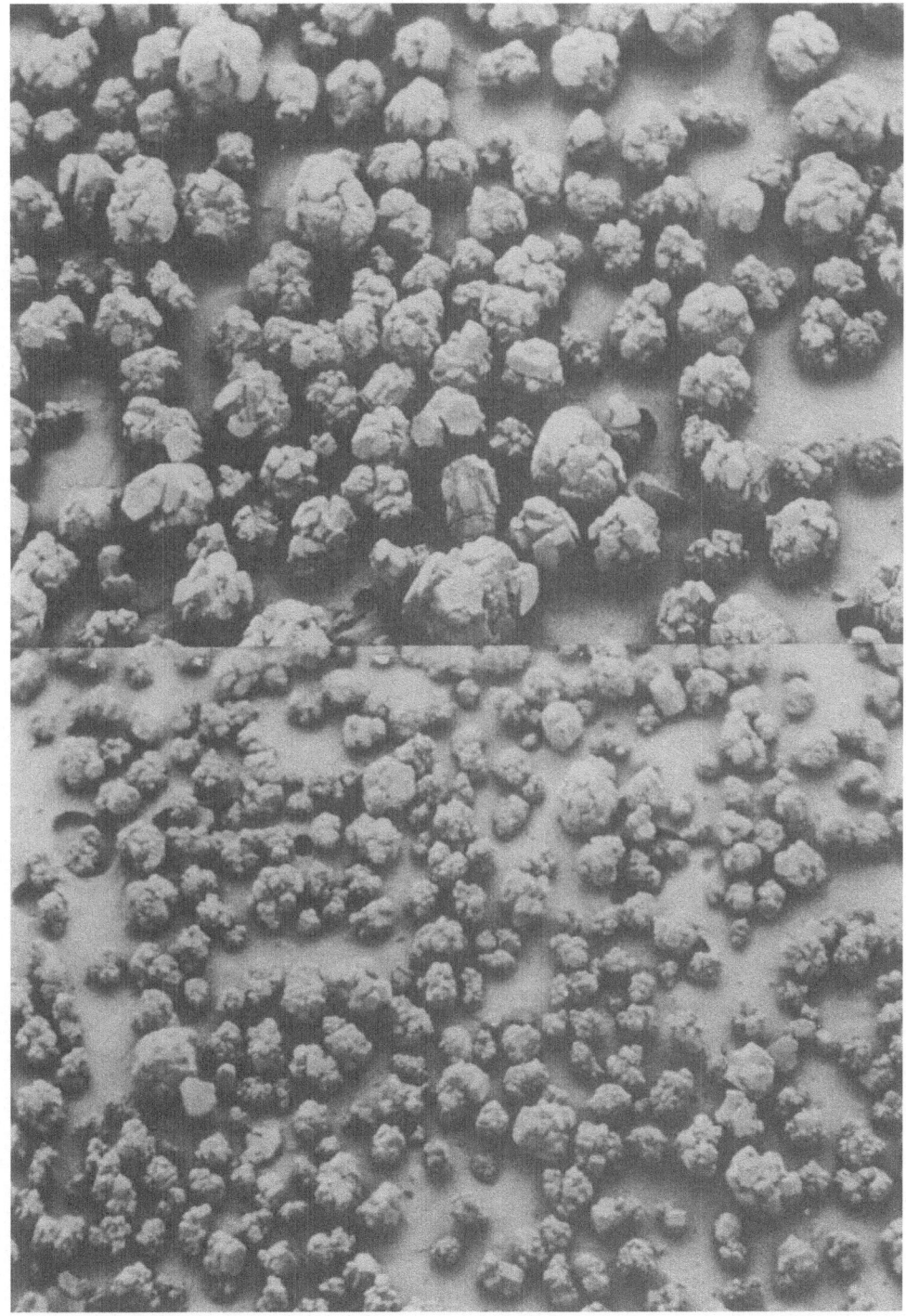

#NC(A)-4

A Note on

ASSESSMENT OF RADIOACTIVITY MEASUREMENTS ON HPLC EFFLUENTS

B.F.H. Drenth, T. Jagersma, F. Overzet,
R.T. Ghijsen and R.A. de Zeeuw

Department of Toxicology
Laboratory for Pharmaceutical and Analytical Chemistry
State University
Ant. Deusinglaan 2
9713 AW Groningen
The Netherlands

In establishing drug metabolite profiles, preferably with radio-labelled parent drug, HPLC is an effective approach, aided by a solvent gradient where the compounds are of widely different character [1]. Detection results can be misleading if obtained off-line on collected fractions rather than on-line: much information is lost if the fraction size is not small enough (Fig. 1). According to Huber et al. [2] the ratio of the fraction size to the S.D. of the peak volume should be less than 0.5, to minimize the possible influence of collection regime. With small fractions, however, the long counting periods needed for high sensitivity and precision [3] delay the procurement of the chromatographic results. Off-line counting is also disadvantageous in respect of time and materials costs.

On-line radioactivity detection, with a flow-through cell, can entail a heterogeneous system, i.e. use of a solid scintillator, or a homogeneous system, i.e. with post-column addition of scintillation fluid to the column eluate. For both methods the detector calibration function takes the form:

$$C = \frac{E \cdot V_d}{F} \cdot A + B$$

where E = efficiency, V_d = effective detector vol., F = flow-rate in the detector, A = activity in the detector cell, B = background counts, and C = gross counts.

Evidently the detector sensitivity depends on the ratio V_d/F. V_d and F cannot be changed at will: decreasing F results in an increased total analysis time, and increasing V_d reduces the (apparent)

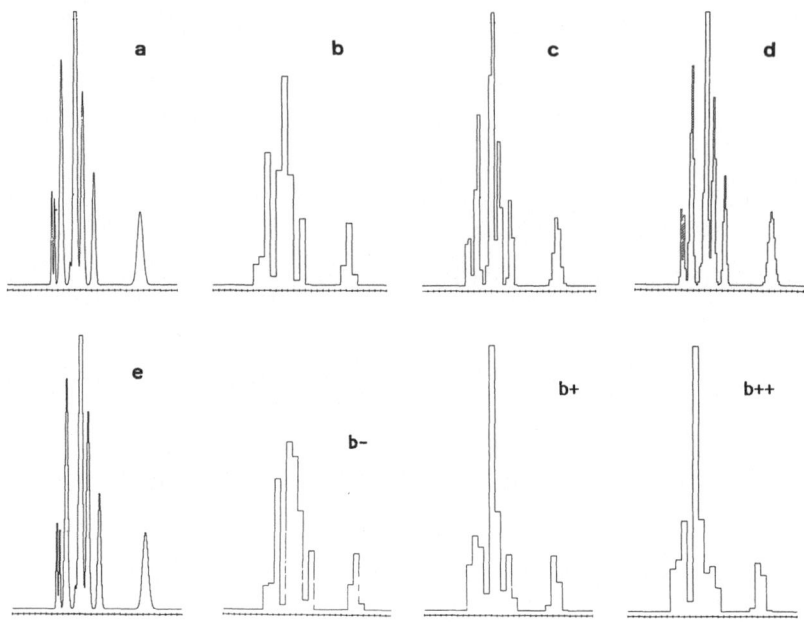

Fig. 1. Influence of fraction-collection settings on information obtained, shown by notional radioactivity results from a chromatogram where peaks were eluted with altogether 30 ml.
a: Pattern from on-line measurement.
Other diagrams represent off-line measurements on collected fractions starting, in the case of **b** to **e**, from the moment of injection (dead volume being disregarded) but with different fraction sizes, viz. **b**, 1.0 ml; **c**, 0.5 ml; **d**, 0.25 ml; **e**, 0.1 ml.
With 1.0 ml fractions as in **b**, the influence of an earlier or later starting point is shown: **b−**, −0.125 ml (i.e. pre-injection collection of 0.125 ml); **b+**, 0.375 ml; **b++**, 0.5 ml onwards.

resolution [3]. Heterogeneous systems are usually much less efficient than homogeneous systems; but with the recent use of small-particle scintillators, comparable efficiency is said to be achieved [4].

Although radioactive detection is very selective, interfering peaks can arise from radioactive contaminants in the solution loaded, such that the material given to the animal must be radiochemically pure, lest impurities (or metabolites thereof) produce peaks. Also, errors could arise from possible isotopic effects both in metabolic and in chromatographic processes.

INTERPRETATION OF SMALL RADIOACTIVITY PEAKS

The question of background signals is now considered, in relation to deciding whether small as distinct from large peaks are genuine.

A Gaussian frequency distribution for background signals (Fig. 2) is usually found, characterized by its mean μ_B and S.D. σ_B. The risk α that a signal will be larger than $\mu + k_\alpha \sigma_B$ can be found from statistical tables, using an appropriate value for k_α. Alternatively, by selecting a maximum risk α, a value for k_α is found which, together with σ_B, gives the limiting value $L_D = \mu_B + k_\alpha\sigma_B$ for the signal. If the risk is R%, then R signals larger than L_D may be expected out of 100 observations. It should be noted that a single-sided limit is used, as we are interested only in signals larger than a certain level: if $>L_D$, the hypothesis that this signal is caused by the background is rejected, and the alternative hypothesis accepted that a peak is present. L_D is thus called the detection limit.

With low-level radioactivity measurements the same approach can be used, but with a Poisson rather than a normal distribution, as in the following example. Suppose that a total count of 6593 has been found by recording background signals for 2 h. For a single counts collection of 10 s, the mean value μ_B for the background is 9.156 counts. The probability $p(n)$ of obtaining an actual signal of n counts in any one counting period can be calculated from:

$$p = \frac{\mu_B^n \cdot \exp(\mu_B)}{n!} \quad .$$

The risk R of obtaining a signal larger than n counts is given by:

$$R(n) = 1 - \sum_{m=0}^{n} p(n) \quad .$$

Table 1 gives some values. Evidently a gross signal larger than 16 counts may be expected 13 times when 1000 measurements are made. Thus in a blank isocratic chromatographic run of duration 1000 × 10 s (i.e. 1000 recordings of the background count), a signal larger than 16 counts can be expected 13 times; hence if L_D = 17 counts, 13 peaks would be detected although no radioactive sample had been injected.

Fig. 2. Gaussian frequency distribution for background signals with mean μ_B: α indicates the risk of observing a signal larger than L_D.

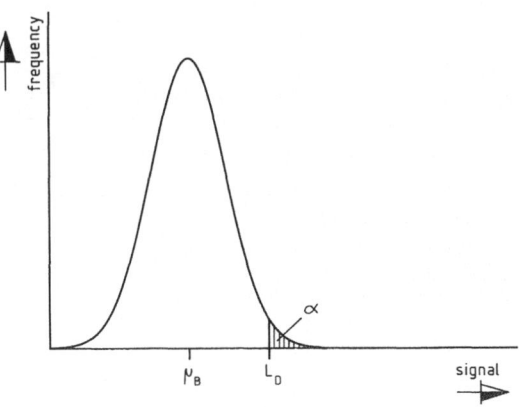

Table 1. Poisson distribution for $\mu_B = 6593/2 \times 60 \times 6 = 9.156$.
The probability of obtaining a signal of n, $\leq$n and >n counts is
denoted p(n), P(n) and R(n) respectively.

n	p(n)	P(n)	R(n) = 1 - P(n)
0	0.0001	0.0001	0.9999
⋮	⋮	⋮	⋮
9	0.1316	0.5667	0.4333
⋮	⋮	⋮	⋮
13	0.0539	0.9180	0.0820
14	0.0353	0.9532	0.0468
15	0.0215	0.9748	0.0252
16	0.0123	0.9871	0.0129
17	0.0066	0.9937	0.0063
18	0.0034	0.9971	0.0029

Any definite peak should be checked for homogeneity, suitably
by NMR, MS, and the application of other chromatographic systems.
The approach described furnishes metabolite profiles that serve for
various comparisons, e.g. different post-dose times, different sample
types (e.g. bile and urine), or different species. If a peak is detected in one particular sample and not another, it does not follow
that the metabolite is present in the latter sample in amounts below
L_D/S' (S' being the overall sensitivity for the analytical procedure).
The frequency distribution of the signal caused by a certain amount
of radioactivity also has to be considered. Fig. 3 shows the general
case. If the distribution of the net signal has a mean value $\mu_S = L_D$,
it can be seen immediately that the risk β of obtaining a signal less
than L_D amounts to 50%!

A mean signal larger than L_D entails a smaller risk. The maximum risk β that can be tolerated results in a limit L_I, the identification limit for the signal (Fig. 4). Fig. 5 summarizes possible
situations.

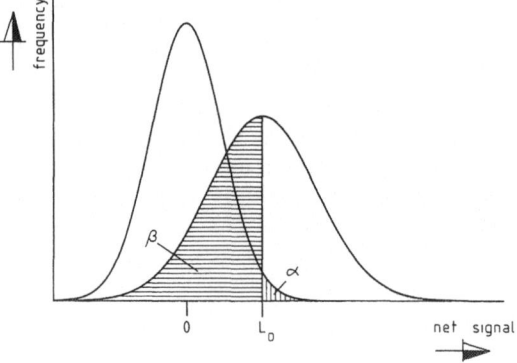

Fig. 3. Gaussian frequency
distribution for background
(left) and true sample *(right)*
with $\mu_B = 0$ and $\mu_S = L_D$.
β indicates the risk that
a sample signal $<L_D$ may be
observed.

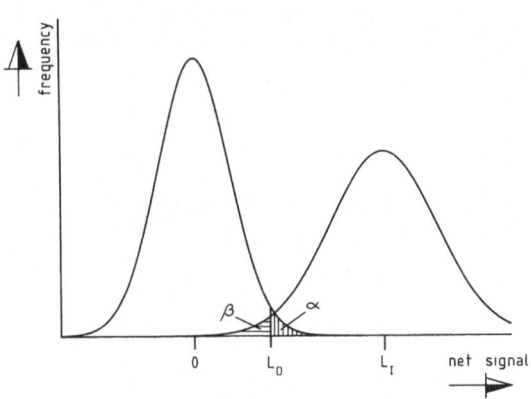

Fig. 4. Gaussian frequency distribution for background *(left)* and true sample *(right)* with $\mu_B = 0$ and $\mu_S = L_I$. See also Fig. 3.
σ_B is the S.D. of the background signal, and σ_P the S.D. of the signal of a true peak with mean μ_S.

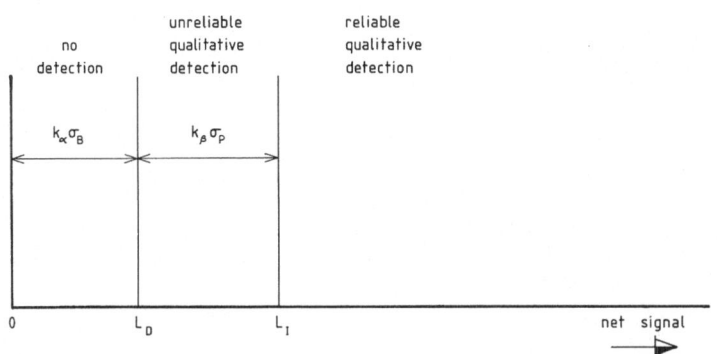

Fig. 5. Conclusions that can be drawn from signals. For amplification see text. *After Currie [6].*

Pursuing the previous example, it may be asked how large an L_I value should be selected in order to keep below 5% the risk of failing to detect a real peak. Table 2 shows that the risk is smaller than 5% if an amount is present that gives a mean signal of 24 counts.

Evidently the detection limit L_D and the identification limit L_I have to be used in quite different situations. The former is used to decide *a posteriori* if something other than background is present; the latter indicates the smallest signal that must be present in order to be sure *(a priori)* that some particular compound will indeed be detected.

For more extensive treatments on detection limits the reader is referred to Boumans [5] and Currie [6].

Table 2. Poisson distribution for several values of μ: n = 17 is the minimum count (L_D) that is not attributed to the background. The probability of obtaining a signal of 17, $\leq$16 and $\geq$17 counts is denoted p(17), P(16) and R(16) respectively.

n	p(17)	P(16)	R(16)
17	0.0963	0.4677	0.5323
$\vdots$	$\vdots$	$\vdots$	$\vdots$
22	0.0520	0.1170	0.8830
23	0.0407	0.0821	0.9179
24	0.0309	0.0563	0.9437
25	0.0227	0.0377	0.9623
26	0.0163	0.0248	0.9752
27	0.0114	0.0160	0.9840
28	0.0078	0.0101	0.9899
29	0.0052	0.0063	0.9937

References

1. Overzet, F., Drenth, B.F.H. & de Zeeuw, R.A. (1981) *J. High Resol. Chromatog.* 4, 448-453.
2. Huber, J.F.K., van Urk-Schoen, A.M. & Sieswerda, G.B. (1973) *Frez. Z. Anal. Chem. 264*, 257-266.
3. Sieswerda, G.B., Poppe, H. & Huber, J.F.K. (1975) *Anal. Chim. Acta 78*, 343-358.
4. Mackey, L.N., Rodriguez, P.A. & Schroeder, F.B (1981) *J. Chromatog. 208*, 1-8.
5. Boumans, P.W.J.M. (1978) *Spectrochim. Acta 33B*, 625-634.
6. Currie, L.A. (1968) *Anal. Chem. 40*, 586-593.

#NC(A)-5

A Note on

QUANTITATION OF ^{14}C RADIOACTIVITY IN HPLC ELUATES

D.W. Roberts

Drug Metabolism Section
Safety of Medicines Department
ICI Pharmaceuticals Division
Mereside, Alderley Park
Macclesfield SK10 4TG, U.K.

The main advantage associated with the use of radiolabelled compounds in HPLC, as extensively practised in our Section's studies of potential drugs, is that a simple non-specific method of detection applies to all labelled components derived from the drug without interference from sample-associated impurities.

The most widespread technique for quantitating radioactivity in the eluate from an analytical HPLC run is the collection of sequential fractions directly into scintillation vials, followed by liquid scintillation counting (LSC). Chromatograms containing low activities can thereby be quantitated using extended counting periods if required. Eluates from preparative separations are collected into tubes, aliquots taken and LSC performed. Fractions containing the required component are then combined and either re-run or following various clean-up procedures submitted for identification etc.

With both of these collection approaches the work involved is laborious and inefficient, and no automation is possible. With the introduction of radioactivity monitors it is now possible to detect ^{14}C components in HPLC eluates without further treatment. The flow-cell, filled with glass scintillator beads (Laboratory Impex), is placed between two photomultiplier tubes using a reflector to ensure the highest counting efficiency. The pulses from the photomultiplier tubes are measured in coincidence to eliminate noise, and chemiluminescence subtraction circuitry and lead shielding reduce the overall background. The radioactivity monitor is ideal for qualitative assessment of ^{14}C patterns or for preparative HPLC where peaks can be detected using the ratemeter and analogue integrator (Fig. 1). The monitor is commercially available (Berthold), as is the ICI interface mentioned below (enquiries to author).

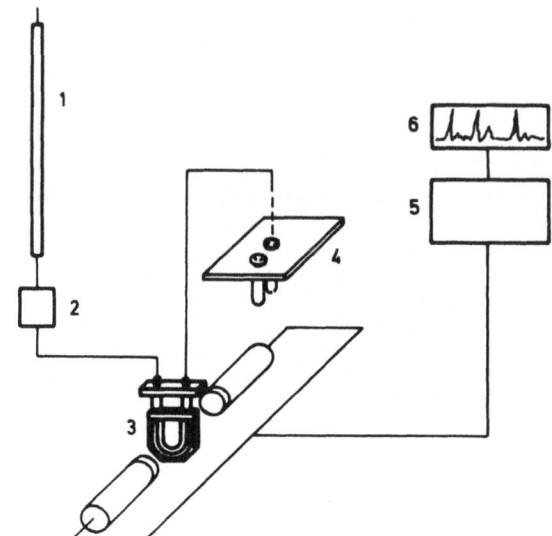

Fig. 1. Schematic representation of glass–scintillator system. **1**, column; **2**, UV monitor; **3**, flow cell with glass scintillator filling; **4**, fraction collector; **5**, electronics; **6**, recorder.

For quantitation of chromatograms the outputs from this monitor are often less convenient. To overcome this at ICI we have interfaced a Commodore PET computer to the monitor and developed a programme with the following requirements in mind: (1) elimination of as much manual processing of data as possible by on-line computation; (2) provision of a wide range of facilities, giving high flexibility combined with simplicity; (3) statistical assesssment of results where appropriate to avoid unreliable or meaningless results being taken at their face value; (4) introduction of autoinjectors etc. to facilitate complete automation.

The input routine starts with questions to the user concerning the sample, HPLC conditions, flow-cell, fraction time, etc. The run is then initiated by the operator, and the digital output from the radioactivity monitor is sampled for the entire run using the fraction time previously entered. Then 10 backgrounds are collected, and all the information transferred onto tape.

The background data are screened for atypical values ('outliers'). If any are found they are rejected by a method dependent on the S.D. and the difference from the average value, and a new average and S.D. are calculated. The minimum test-sample % value that will differ significantly from zero is then calculated. The data from the run are analyzed for statistical significance in relation to the backgrounds, and processed to determine peak location, retention time (here termed R_t), % composition and confidence limits. Fig. 2 shows the outcome, with identifying headings for the particular run.

Several variables affect the monitor output: (1) counts injec-

```
INDEX NO      : 1/M3                COMPOUND : ICI/123456
PROTOCOL NO : ICI/123456/01         STUDY NO : 1        EXPT NO : 1
USER        : D ROBERTS             DATE     : 23/06/81

MONITOR NUMBER       : 3            FLOW CELL VOL (UL)  : 200
RUN TIME (MIN)       : 10           FLOW RATE (ML/MIN)  : 1
FRACTION TIME (MIN): 1              CELL  RESIDENCE (MIN): .2
SAMPLING TIME (MIN): .5
SOLVENT SYSTEM       : 65% ACETONITRILE - 35% WATER
COLUMN DETAILS       : 5 UM SPERISORB ODS
OTHER DETAILS        : NONE

SAMPLE ID    : 100 UL TEST SAMPLE (PRECISION TEST)
             *********************************
             RETENTION TIME (MINS)
0.0    2.00    4.00    6.00    8.00    10.0
```

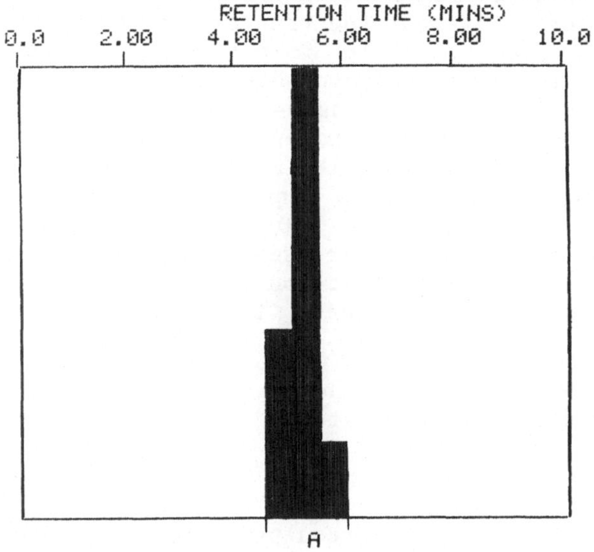

```
------------------------------------------------------------------
I PEAK I    FRACTION NO.    I  RT  I % OF I CONF. LIMITS   I         I
I REF. I START  PEAK  STOP  I(MINS)I TOTAL I (% : 2P=0.05) I COMMENTS I
I ---- I -----  ----  ----  I ---- I ----- I ------------- I -------- I
I  A   I   6     6     7    I  11  I 100.0 I 96.7 TO 103.3 I         I
------------------------------------------------------------------
TOTAL NO OF PEAKS = 1 (TABULATED ONLY IF >1%)

NET HISTOGRAM TOTAL = 1103.99 COUNTS (CONF. LIMITS = +/- 66.05 COUNTS)
APPROX SAMPLE DPM   = 27599 (FOR 20 % STATIC COUNTING EFFICIENCY)

BGDS: 5 10 4 4 13 5 10 11 2 10

AV BGD = 7.4 +/- 3.77 COUNTS (SD : N= 10 )
MSP    = .81 % ( 16.34 COUNTS : '.....' LINE)

COMMENTS :

SIGNATURE :
DATE       23/06/81
```

Fig. 2. An illustrative print-out

ted; (2) no. of components in chromatogram; (3) fraction interval;
(4) flow rate; (5) flow-cell size. Various tests related to these
variables were performed with a ^{14}C drug to compare the monitor with
computer data assessment and conventional fraction collection with
LSC. Generally the HPLC conditions (Fig. 2) were optimal. A nominally
200 µl flow cell was used in the monitor.

(a) ASSESSMENT OF THE MINIMUM INJECTABLE DISINTEGRATIONS/MIN

When an estimate of some quantity has been made, one wants to
know its confidence limits, maybe set very high, e.g. 99% certainty,
or more modestly at 80, 90 or 95%. For radiochemical purity analyses a confidence range value of <1% about the true result is needed;
for routine analytical work a range of up to 5% is acceptable. For
preparative work a much wider range may be used, depending on requirements. Table 1 shows the analysis of a single-component standard
containing 1,000–140,000 dpm, aimed at determining the limit for a
given amount of radioactivity applied. With the monitor and PET computer, ≮69,000 dpm in a single component is required for a confidence range of ±1%, or 7,000–14,000 dpm for ±10%. Below 7,000 dpm
applied the confidence range dramatically increases to a point where
the result is totally meaningless. With fraction collection ≮7,000
dpm is needed for a ±1% range, while ±10% needs <1,400 dpm.

The low number of counts detected by the flow-cell when compared with the counts derived by LSC is the major problem of a
dynamic system such as the radioactivity monitor.

(b) PRECISION OF FLOW-CELL/COMPUTER DATA

Table 2 shows that flow-cell/computer results agree well with
fraction collector/LSC results although the spread about the mean is
larger, viz. S.D. ±2.8% compared with only ±0.08%. The confidence
range for the flow-cell data is also higher than in the fraction-
collector method, due to the lower no. of counts detected.

(c) EFFECT OF CHANGING THE FRACTION INTERVAL

The equation used to determine the confidence range of a peak
in a histogram depends on (1) the no. of fractions making up the
peak; (2) the no. of fractions in the whole histogram; (3) the no.
of fractions used to determine the mean blank value; (4) the gross
counts for the peak in question and for the entire histogram. On decreasing the fraction interval to obtain greater resolution, the
confidence range must widen due to the increase in no. of fractions.

Table 3 shows that with fraction times ranging from 1 down to
0.1 min the total no. of counts detected and the peak % for each run
were very similar. At a sampling interval of 0.4 min the confidence
range started to increase, and at 0.2 min it was >±5% threshold.

Table 1. Influence of radioactivity load: Radioactivity Monitor/PET Computer compared with Fraction collection (0.5 min intervals)/LSC. The flow conditions were as in Fig. 2. 'Peak %' values are relative to total cpm detected, and 'c.r.' signifies the confidence range as % for $2P = 0.05$; 'cpm, obs.' = counts detected.

dpm in-jected x 10^{-3}	Monitor + Computer		Fraction collection + LSC	
	cpm, obs.	Peak % ±c.r.	cpm, obs.	Peak % ±c.r.
139	5687	100.20 ±0.8	115408	99.22 ±0.08
69	2868	98.86 ±1.34	55944	98.99 ±0.16
27.7	1104	100.00 ±3.3	22039	99.24 ±0.36
13.9	512	102.30 ±7.6	11303	97.65 ±0.71
6.9	266	102.80 ±13.6	5477	99.36 ±1.39
2.8	126	92.22 ±22.98	1986	99.26 ±4.4
1.4	62	93.55 ±48.15	1025	97.03 ±7.09

Table 2. Precision assessment: comparison as in Table 1 (dpm injected = 27.71×10^3). Values are Peak % ± confidence range.

Run no.	Monitor + Computer	Fraction collection + LSC
1	97.04 ±3.36	99.03 ±0.36
2	100.0 ± 3.30	99.24 ±0.36
3	100.10 ±2.9	99.18 ±0.35
4	103.00 ±4.3	99.22 ±0.35
5	97.35 ±3.15	99.16 ±0.37
6	100.00 ±3.10	99.09 ±0.37
Mean	99.58	99.15
±S.D.	2.18	0.079
C.V.,%	2.19	0.08

Table 3. Effect of varying the fraction time on results as obtained by Monitor/Computer. See heading to Table 1; dpm = 22,231.

Time, min	cpm, obs. (net)	Peak % ±c.r.
1	728	96 ±3.07
0.9	648	97 ±3.7
0.8	688	97 ±3.9
0.7	728	104 ±3.2
0.6	687	102 ±3.9
0.5	771	97 ±3.4
0.4	674	99 ±4.4
0.3	674	100 ±4.7
0.2	707	102 ±6.0
0.1	722	102 ±8.7

(d) EFFECT OF CHANGING THE FLOW RATE

The actual number of counts detected by the radioactivity monitor is affected by two major criteria, (1) flow-cell size, and (2) the flow rate of the eluate through the cell, which if combined determine the residence time of the ^{14}C in the flow-cell. Table 4 shows that the main effect of increasing the flow rate is to proportionately decrease the no. of counts detected, i.e. the residence

Table 4. Effect of varying the flow rate on results as obtained by Monitor/Computer. For 'Peak % ± confidence range' see Table 1 heading; same fraction time (0.5 min) and flow-cell size (200 µl). The ^{14}C injection was 22,231 dpm.

Flow rate, ml/min	cpm, obs.	Peak % ± confidence range
0.5	1471	100 ±3.0
1.0	771	97 ±3.4
1.5	453	105 ±5.8
2.0	316	104 ±10.3

Table 5. Analysis of 9 replicates of an incubation mixture (with auto-injection) using the Radioactivity Monitor and PET Computer, with a 200µl flow cell. The total radioactivity loaded represents 100% (cf. heading to Table 1); ↑ = maximum peak %, ↓ = minimum peak %.

Sample	Peak % ± confidence limits (maximum + minimum)		
	Metabolite 1	Metabolite 2	Parent compound
1	Not resolved at 0.5 min/ml fraction (9.03 ±0.47)		88.97 ±0.74
2–9	↑ 1.76 ±0.21	↑ 7.40 ±0.42	↑ 88.21 ±1.13
	↓ 1.03 ±0.16	↓ 6.04 ±0.36	↓ 85.08 ±0.78
Mean %	1.43	6.71	86.92
± S.D.	0.20	0.46	1.03

time of the ^{14}C in the eluates is proportional to the flow rate. This decrease in counts also widens the confidence range.

The residence time can be increased by using larger flow-cells, e.g. 400 µl; but a major disadvantage in doing this is that the peak shape may broaden and so reduce the resolution available.

(e) A TYPICAL ANALYSIS OF AN INCUBATION SAMPLE USING AUTO-INJECTION

Ten replicate samples were analyzed by RP-HPLC to determine the concentrations of two metabolites in a microsomal incubation mixture. Table 5 shows the proportions of the two metabolites and parent compound as % of the total calculated by the computer. The metabolites, structurally very similar, are only just separated by the HPLC system developed, and cannot be resolved using 0.5 min fractions (see run 1).

By using 0.25 min fractions the two metabolites can be resolved and quantified. This fraction time would be very difficult using fraction collection. The data demonstrate that even minor components within a separation can be quantified with statistical confidence provided that sufficient radioactivity is loaded.

DISCUSSION

The quantitation of ^{14}C eluates by on-line radioactivity monitoring has proved very successful in our laboratories. By combining the simplicity of the radioactivity monitor and the flexibility of the PET computer we have developed a system which is simple to operate, performs on-line data processing with statistical interpretation, and allows the use of auto-injectors.

We have identified three major disadvantages with the system, notably the low sensitivity of the actual flow-cell (about 20-30%). This has to some extent been improved on by the introduction of a new flow-cell packing which has counting efficiencies in excess of 50%; but this packing is more susceptible to increased background (memory effect) and because the particle size of the packing is much smaller, the flow-cell back pressure is correspondingly higher, which prevents flow rates above 2 ml/min being used.

Sample types which we are unable to analyze using the radioactivity monitor are typically plasmas corresponding to a long post-dose interval, some urines, tissue extracts, etc., i.e. samples of low activity.

The second disadvantage which we have identified is associated with the collection of background data. Currently we collect 10 values but often find that when the counts in the histograms are low the appearance of peaks unrelated to ^{14}C increases, i.e. the actual variation in the background value becomes very important. One possible solution to this problem could be the collection of a complete blank run.

The third disadvantage is associated with the analysis of the histogram, especially when two or more components are not fully resolved by the HPLC solvent system. This problem is not easily solved since the computer is working to fixed criteria which do not always apply to peak valleys and tails etc.

All the above disadvantages increase in severity as the number of counts applied to the column decreases.

The major advantages over fraction collection are quite numerous. We have found the system to be very easy to operate and have successfully joined on an auto-injector which allows HPLC analyses to be done overnight or during the weekend. With fraction collection much time is devoted to filling and capping vials which then have to be counted, or with preparative work further aliquotting is necessary which not only depletes the amount of valuable sample but with very long run times can be very time-consuming.

With the radioactivity monitor/computer system the raw data are processed immediately with full statistical interpretation of the results; therefore the time required in the development of methods for analytical and preparative purposes is markedly reduced.

We have found the system very flexible, readily allowing change of flow rate, fraction time and flow-cell to obtain the best resolution for a particular sample. It is also possible to do gradient-elution HPLC without the problems associated with the counting of complex solvent mixtures by LSC. Finally, savings in scintillants and vials are easily achieved, whilst the problems associated with the disposal of large volumes of scintillant from HPLC analyses are eliminated.

Relevant literature

1. Mackey, L.N., Rodriguez, P.A. & Schroeder, F.B. (1981) *J. Chromatog.* *208*, 1-8.
2. Reeve, D.R., Yokota, T., Nash, L.S. & Crozier, A. (1976) *J. Exp. Bot.* *27*, 1243-1258.
3. Webster, H.K. & Whaun, J.M. (1981) *J. Chromatog.* *209*, 283-292.

Comments on material in #A

Comments on #A-2, W. Dieterle & J.W. Faigle – PREPARATIVE HPLC

W. Dieterle, *answering* C.R. Jones.– For the 25 mm i.d. steel columns our flow-rate is 200 ml/min; we do not measure efficiency, but judge each column by whether it achieves the desired separations. *Answering* P. Hendley.– The high-resolution large-scale columns last for ~50 injections provided that their expensive packings are protected by guard columns which contain the same packing material and whose lifetime is <5 injections. Pre-purification procedures are conducted using cheaper packing materials with relatively short lifetimes. *Queries by* H.J. Egger.– Sjövall and his group have made extensive use of lipophilic Sephadex derivatives, e.g. Lipidex 5000, to separate mono- and di-glucuronides and other metabolites: have you tried this ? [*Reply:* No!] – With XAD-2 have you observed any adsorptive losses such as occur with dextran gels? *Reply.*– Irreversible adsorption has never been seen with micronized XAD-2 in prep-LC under the conditions described (in #A-2). Generally (*reply to* P. Hendley) none of our techniques are prone to artefacts, as routinely looked for in our isolations.

In reply to J. Chamberlain and H. de Bree.– Before structural analysis is performed on metabolites separated by prep-LC, further clean-up is done, always by HPLC. Whatever the technique to be used finally, we aim to produce pure material.

Questions by U.A.Th. Brinkman.– (1) Is 12 μm XAD-2 commercially available, and in a narrow-range fraction ? (2) How does it compare in performance with, e.g. 10 μm LiChrosorb RP material ? (3) Is there any special reason for including 1,2-dichloroethane in the mobile phase for XAD-2 but not for LiChrosorb RP? *Editor (the authors' replies not being at hand).*– The position concerning (1) at the time the relevant paper (ref. [18] in #A-2) was published evidently entailed 'in-house' toils to prepare the micronized XAD-2.

Comments on #A-3, C.R. Jones – PROBLEMS WITH LOW CONCENTRATIONS

C.R. Jones, *replying to* J. Ramsey, in respect of contaminants in nitrogen after conduction through PVC tubing. – Whilst the GC peak was shown to be due to tributyl aconitate, this was not established as the cause of the fluorescence peak in HPLC.   *Added by Editor* (E.R.).- PVC (but not PTFE) tubing has also been reported to give a rogue GC-ECD peak in the oxprenolol position, due to bis(2-ethyl-hexyl)phthalate [1]. C.R. Jones, *answering* M. Bonafé, who mentioned the effectiveness of preparative columns (e.g. Lobar from Merck) in giving water free of HPLC peaks.- We have obtained high-quality water similarly, with mixed RP material packed in our laboratory; the main problem with pure water is how to keep it so.   *Comment from* B.-A. Persson, on filtering the mobile phases used in gradient elution: UV-absorbing impurities are dissolved from Millipore filters even with a very low content of methanol or acetonitrile. *Answer by* C.R.Jones - We have chosen to use glass-sintered filters because they are cheaper although less efficient.

Comments on solvent re-cycling in HPLC.- (C.R. Jones, *responding to* D. Dell, I.D. Wilson) We can re-cycle for up to several weeks, depending on the separation.(D. Dell) In our laboratory experience has been good: the stability of the system seems to improve with time, and the standing (background) current does not increase during several days. However (B. Scales) re-cycling can give problems in RP systems using fluorescence detectors: endogenous biological components which accumulate in the solvent can polymerize under the influence of the intense UV light and cause severe clouding of the windows.

1. Hooper, W.D. & Smith, M.T. (1981) *J. Pharm. Sci. 70*, 346-347.

Comments on usefulness of an internal standard (i.s.) in HPLC

Note by Editor (E.R.).- A discussion at the Forum indicated that for HPLC, as for GC [repeatedly considered in earlier books - e.g. by J. Vessman and S.H. Curry in Vol. 10, #NC(F)-2 & #NC(F)-3], use of an i.s. is variously regarded as an asset to precision and as a dubious help even where the chosen compound is shown to match the analyte in behaviour relevant to work-up losses. A compound dissimilar in this respect and therefore inapplicable as a 'processed i.s.' could serve as a 'load standard' at the chromatography stage [terminology advocated by E. Reid (#A-6 in Vol. 7); some authors use the term i.s. merely in the latter sense]; but even this practice may be questionable, as the following précis makes evident.

Some views expressed.- Only where an automatic injector is used might the Glasgow group (M.J. Stewart, I.D. Watson) feel confident without an i.s., and routinely they do use a processed i.s. to obviate possibly serious imprecision associated with sample work-up. B.-A. Persson likewise feels an i.s. is of benefit to pre ision. C.R. Jones and colleagues usually employ an i.s.; yet with low-concentration samples precision can suffer, merely because normal experimental error gives peak-size variations from one injection to another; a 20% rise in sample peak with a 20% fall in i.s. peak when a replicate is analyzed would raise the apparent concentration by 50% with the usual ratio-based calculation. D. Dell was also cautionary about expecting improved precision with an i.s.; a check is needed, as advocated by P. Haefelfinger (Avignon meeting, 1981).

Ion-pair HPLC approaches - B.-A. Persson (AB Hässle, Mölndal)

Editor's abridgement of a post-Forum talk (cf. G. Schill in Vol. 7, #D-1).- With a bonded phase, and the agent in the mobile phase, the retention data may suggest either ion-exchange or ion-pair adsorption. Organic modifiers, e.g. aliphatic amines where hydrophobic amines have to be separated, may improve HPLC performance and can affect retentions. Bonded phases can be used in the liquid-liquid mode, but this is usually done with silica (NP) onto which an aqueous stationary phase is coated: an immiscible organic phase elutes the ionic compounds paired with a counter-ion which can be chosen to confer detectability [but see comment in #E - *Ed.*]. A rather neglected variant is ion-pair adsorption, e.g. for amines in biological samples: the mobile phase contains, e.g., methanol and aqueous perchloric acid in minor amounts as additives to dichloromethane.

Comments and points related to #A-4, M.J. Stewart - EXTRACTION, etc.

Reply to J. Tomašić.- With acidic metabolites, of which we have less experience than with basic metabolites, the main problem is the large number of endogenous acids in urine, as evidenced by 'control' chromatograms. Possible approaches are improved extraction selectivity, maybe with ion-pairing, and more specific detection; the whole assay procedure may have to be tailored to the analyte of interest. [See comments on #A-5, below.-*Ed.*]

Discussion points.- L.E. Martin and others had encountered the problem of drug binding to ultrafiltration membranes. In principle, however, ultrafiltration is reckoned by M.J. Stewart to be an effective way of getting values for 'free' drug, which (as M.E.J. Vince remarked) is generally believed to relate more closely than 'total' drug to clinical effects. M.J. Stewart instanced thiopentone as an exception to this generalization, its CNS level being 5 times 'total' plasma level and much greater than the 'free' level, and stressed the need in the ultrafiltration approach, if the 'free' drug level is indeed wanted, to start by ascertaining adsorptive losses with spiked aqueous samples. Plasma-protein removal must be rapid, with pH control of the filtrate to compensate for the removal of protein anions. The sensitivity of the end-reaction may of course have to be increased. It may be hard to find an i.s. with ideal behaviour.

Answering M.J. Stewart, U.A. Th. Brinkman outlined the scope of 'trace enrichment' in relation to pre-concentrating large volumes for HPLC, in the 1-10 ml or 100-1000 ml range. The short (2-5 mm) pre-columns used, best packed with the same material as in the main column, are quick to load and readily interchangeable. Problems may occur with very dirty samples, when diverse compounds in large amount may 'over-load' the pre-column and separation may be inadequate when it is subjected to gradient elution. Then, if selective detection does not help, one will have to resort to an additional clean-up step. [For amplification, see #B-3 in Vol. 10.- *Ed*.]

Points from a sample-preparation talk by B.-A. Persson.- A direct injection may be feasible with urine, where analyte concentrations are often high and selective detection can help. Plasma usually has to be deproteinized, with the risk of occlusion losses (easily checked, as remarked by D. Dell who felt this approach to be the best one for highly water-soluble compounds such as cephalosporins). The extent of work-up is influenced by sample-volume considerations and column stability. Liquid-liquid extraction is still the general approach where appropriate, with alertness to adsorptive losses and interferences. Back-extraction into a small aqueous phase or, more commonly, evaporation serves to concentrate the extract, or it may be directly injected into an NP column – a rational approach that deserves wider use. *Editorial comment*.- This sketch has counterparts and reinforcement throughout this book (e.g. in #A-2 and #E) and in previous volumes; no terse account can do justice to the 'lore' that has to be learnt, somewhat lessened by the advent of HPLC (see #O in Vol. 10). Acetone deproteinization, albeit less efficient than perchloric acid, may be advantageous for tightly bound drugs, especially if acidic.

Double-derivative spectroscopy in HPLC (reply to C.J. Lewis): A.F. Fell mentioned success with this approach in distinguishing incompletely separated peaks where resolution could not be achieved by conventional means such as changing the mobile phase.

Automatic analyzers with provision for sample handling.- Concerning the DuPont Prep I as transiently marketed in the U.K., L.E. Martin and M.J. Stewart made remarks which tallied with what C.R. Jones said – that the principle of liquid-solid extraction is a good one, that the instrument was well made and performed well, and that its withdrawal was deplorable. He suggested that any new version be based on pneumatics rather than centrifugation, enabling up to 100 samples to be handled. The Technicon Fast-LC system also received favourable mention, notably from L. Abrahamsson and I.D. Cockshott. Their experiences included, respectively, anticonvulsant and tricyclic-antidepressant monitoring, and assays for an acidic drug – which entailed modifications to minimize problems of carry-over, reproducibility, and reliability in unattended running.

Comments on #A-5, B.L. Goodwin - CAPILLARY GC OF PHENOLIC ACIDS

   Reply to J. Chamberlain.- The flow rate at which a change from laminar to turbulent flow is to be expected can be calculated using physical dimensions. Although capillary columns are reckoned to be easily overloaded (*point raised by* A.S. Papadopoulos), the limit in our case may be as high as 5 µg. *Comments by* J.B. Hopper.- The stated need to condition the column each day for quantitative analysis may relate to the injector, not the column, as suggested by our experiences following the advent, in capillary GC, of on-column injection as applied by us to thermally labile compounds and those with a high affinity for adsorptive surfaces. *Reply to* J. Vessman.- The 'Grob-type' injection was splitless, say 2 µl (10 µl might 'strip').

   An editorially contributed ref.- Assays on CSF exemplify the application of   RP-HPLC (C-18) in this field [1]. With a citrate-phosphate buffer and amperometric detection (glassy carbon electrode) 4-hydroxy-3-methoxyphenylglycol (HMPG; cf. Fig. 3 in #A-5) could be determined directly. An ion-pair system also allowed detection of the ethanol analogue and of acids such as VMA and HVA.

1. Langlais, P.J., McEntee, W.J. & Bird, E.D. (1980) *Clin. Chem.* *26*, 786-788.

Comments on colour reactions - cf. #A-6, W. Ritter - HPTLC
                              & later articles including #D-5

   W. Ritter, *replying to* M.J. Stewart who had questioned whether colour reactions are an important assay tool.- An example of a study where a specific colour reaction was indispensable concerned the absorption of imidazole antimycotics through skin or after vaginal application. The high sensitivity attained, 100-fold that by UV measurement, enabled 0.3 ng/ml levels to be determined in plasma, the levels being in fact below 10 ng/ml for drug and metabolite [#D-5].

   Remarks by J.A.F. de Silva.- Colour reactions and fluorescence derivatization (e.g. o-phthalaldehyde for primary amines) can help for various purposes including sensitive post-column detection in HPLC. The Bratton-Marshall reaction for primary aromatic amines is so specific as to be invaluable for identifications. In general, signal-to-noise ratio and hence detection sensitivity can be improved by minimizing the background due to the TLC plate, as in fluorescence or phosphorescence assays where chilling enhances analyte luminescence. *Room temperature phosphorescence (RTP): comment by* A.F. Fell.- Reduction of background has been achieved [1] by transforming the emission densitometric scan output to its 2nd derivative record.

   Editorial note on literature sources.- Guidance on reactions and reagents is to be found in classical books on TLC (E. Stahl, J.G. Kirchner) and on spot tests (F. Feigl) besides the 'Merck Index', and also, for bioconstituents and certain drugs including antibiotics, in ref. [2].

1. Vo-Dinh, T., Gammage, R.B. & Martinez, P.B. (1980) *Anal. Chim. Acta 118*, 313-323.
2. Dawson, R.M.C., Elliott, D.C., Elliott, W.H. & Jones, K.M. (eds.) (1969) *Data for Biochemical Research*, 2nd edn., Clarendon Press, Oxford (see pp. 509-591).

Comments on #**NC(A)**-1, D.A. Stopher – QUANTITATIVE TLC

Response to queries by H. Hamböck.– Possible degradation on the plate during analysis, e.g. photolytically or by esterification in alcohol/acid mixtures, has not been a problem with most of the compounds that we have analyzed, although some have shown photolytic decomposition during scanning on the Vitatron densitometer; none showed breakdown due to solvent interaction. However, when analyzing ng amounts of some compounds we have noticed breakdown (probably oxidation) during application to the TLC plate. Concerning the time required, about $2\frac{1}{2}$ h is needed to prepare 20 plasma samples (i.e. pipette sample, add i.s., extract, take to dryness, and transfer to TLC plate); chromatography takes about 1 h and densitometry 0.75 h.

Comments on #**NC(A)**-4, B.F.H. Drenth et al. – LABELLED HPLC PEAKS

Concerning actual determinations (not featuring in the Note). – *Reply to* W. Dieterle: We can't make efficiency measurements on-line but have to deal with quenching by estimating and interpolating the efficiency from several separate determinations; maybe the channels ratio approach with a dual-channel counter would help. *Answering* H. de Bree: not having tried solid scintillators, we don't know if there would be chemiluminescence problems. *Ref. contributed by Editor.*– In the context of MS on-line with HPLC, a perhaps over-pessimistic view has been expressed concerning the capability of fraction collection to furnish discrete peaks for off-line examination [1].

'Housetraining your PET' [cf. #**NC(A)**-5]

Readers unfamiliar with this computer may consult a lively account of its role in automated instrumentation [2].– *Ed.*

1. Arpino, P.J. & Guiochon, G. (1979) *Anal. Chem. 51*, 683A-701A.
2. Stieg, S. (1982) *Lab. Pract. 30*, 701-705.

Section #B

INVESTIGATION OF METABOLITES, ESPECIALLY PHASE I

#B-1

EXAMPLES OF APPROACHES TO ISOLATING
AND IDENTIFYING PHASE I METABOLITES

G.R. Bourne and S.R. Moss

Drug Kinetics Goup, Safety of Medicines Department
Pharmaceuticals Division
Imperial Chemical Industries PLC
Alderley Park, near Macclesfield, SK10 4TG, U.K.

The Phase I metabolites of ICI 89,406, a β-adrenoreceptor blocking drug, were isolated from bile from the perfused rat liver preparation, and the structures determined by NMR spectroscopy. The metabolic fate of the veterinary product cloprostenol ('Estrumate'*), a prostaglandin analogue, was studied in vivo in the target species, the cow. GC-MS was used to identify drug-derived urinary components that had been isolated by a range of chromatographic procedures. In the study of cloprostenol metabolism the high isotopic abundance of ^{14}C in radiolabelled cloprostenol allowed an unusual method of detecting metabolites.

This article describes studies undertaken with two structurally different drugs in respect of isolating and identifying Phase I metabolites[†]. Radiolabelled compounds as used throughout these investigations provided a sensitive, straightforward means of detecting the drugs and metabolites. The descriptions that follow illustrate the diversity of samples analyzed and methods used in metabolism studies.

(1) ICI 89,406, A β-ADRENORECEPTOR-BLOCKING AGENT

Studies *in vivo* demonstrated that in the rat this compound (see overleaf for structure) was extensively metabolized and that the major route of excretion of drug and metabolites was the bile. Because of the importance of biliary elimination, metabolic studies were done with the isolated perfused rat liver.

---

* a Trademark, the property of Imperial Chemical Industries PLC
[†]For explanation of 'Phase I' (*vs.* 'Phase II' = conjugates) see #A-2.—*Ed.*

^{14}C–ICI 89,406

OCH$_2$CH(OH)CH$_2$NHCH$_2$CH$_2$NHCONH

* denotes the radio-
label position

CN

Isolated liver perfusion preparations were used on 3 occasions, with livers from male rats (200 g body wt.). A 10 mg dose of ^{14}C drug was used *in vitro* on each occasion. Bile was collected for up 3 h after addition of the drug via the reservoir to the circulating perfusate (100 ml; heparinized blood:saline, 2:1).

The mean recovery of the applied dose in the bile was 51.8%. The remainder was recovered in the perfusate and liver, giving a mean total recovery of 88% for the ^{14}C. *In vivo* studies of the fate of the orally dosed drug and a preliminary study with the perfused liver indicated that the drug metabolites present in bile were mainly conjugates. To liberate the Phase I metabolites the 0–90 min bile from each experiment, containing over 90% of the eliminated metabolites, was pooled and treated with β-glucuronidase. TLC analysis demonstrated that this procedure liberated several aglycones.

The sample of bile after hydrolysis was lyophilized. The metabolites were quantitatively recovered from the lyophilized residue in the chromatographic solvent and analyzed by use of a C-18 column with acetonitrile/water/ammonia as the eluent.

Three major metabolites designated 2b, 2cI & 2cIII were isolated and taken for MS-EI and NMR analysis. Their isolation by the above procedures gave 200–400 μg of each, as a white solid. No useful information about the structure of the three metabolites was obtained from MS analysis. NMR analysis was then performed, using deuteromethanol as solvent and with the spectrometer set in the Fourier transform mode. Integration of the aromatic protons and the protons next to the oxygen demonstrated that hydroxylation of ICI 89,406 had occurred. The aromatic regions of the spectra of an authentic sample of ICI 89,406 and the isolated metabolites are shown in Fig. 1.

The protons at 7.55δ in ICI 89,406 (Fig. 1, Spectrum 2) are meta to the oxygen function (H$_3$ & H$_5$). The complex multiplet at 6.8–7.5δ contains the remaining protons. Introduction of a hydroxyl function into a benzene ring causes upfield shifts (i.e. lower δ values). Typical values are: o-proton ≈ 0.5 ppm, *m*-proton ≈ 0.1 ppm, *p*-proton ≈ 0.4 ppm. If hydroxylation had occurred at C$_3$ or C$_5$, one of the protons at 7.55δ would be lost and the other would shift upfield by 0.1 ppm. In addition H$_4$ and H$_6$ would move upfield out of the multiplet 6.8-7.5δ.

The combination of these two predictions was not observed in either 2cI (Aglycone 2, Fig. 1) or 2cIII (Aglycone 3, Fig. 1). In

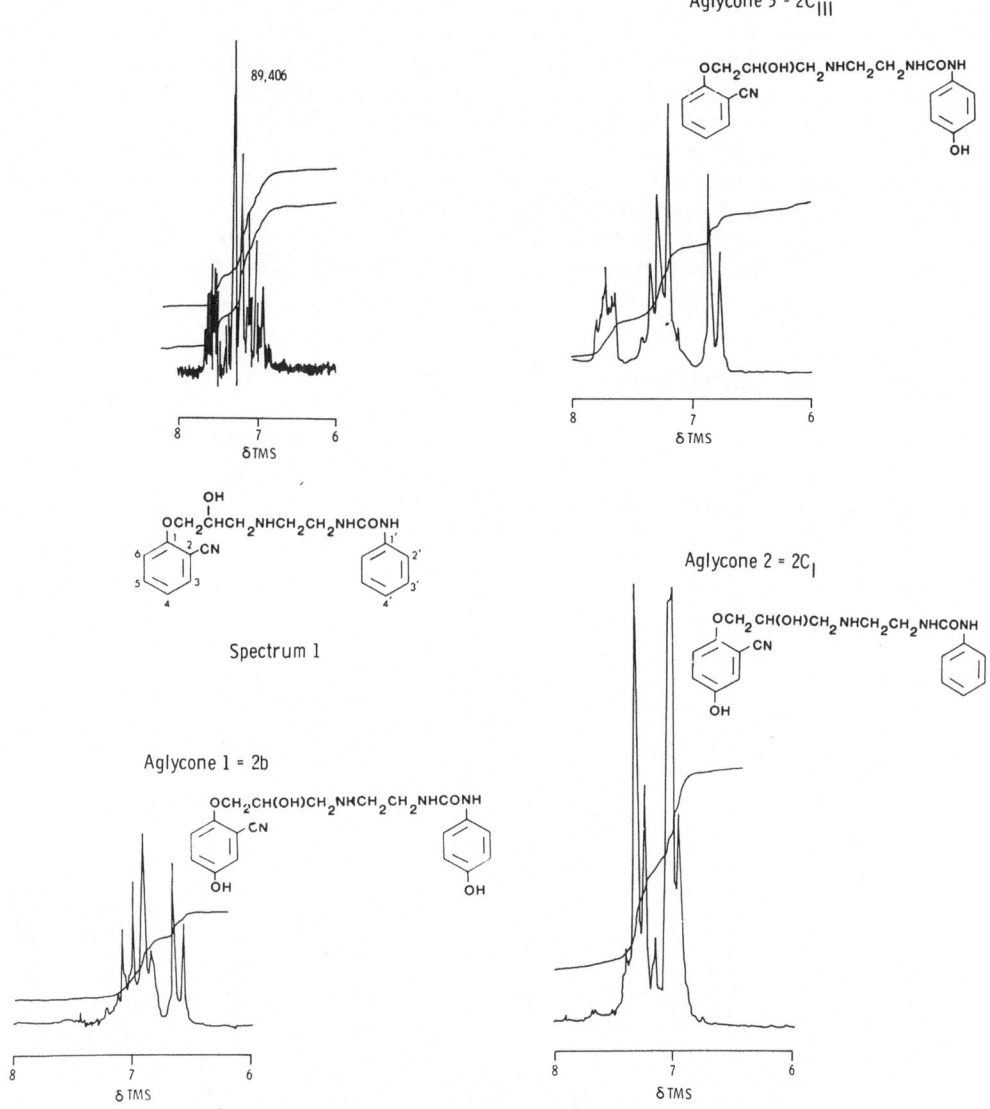

Fig. 1. Partial NMR spectra of ^{14}C-ICI 89,406 and its metabolites.

2cI *both* protons at 7.55δ have moved upfield by ~0.5 ppm into the multiplet at 6.8 to 7.5δ, indicating substitution at C_4 or C_6. Unfortunately this NMR spectrum will not distinguish between these

two substitution positions. 2cIII still retains the protons at 7.55δ
but two protons have appeared at 6.7δ. These are the protons ortho
to the hydroxyl function at C_4, and the doublet pattern is typical
of the pattern for a para-substituted benzene ring. Thus the structure of 2cIII is clearly defined as shown in Fig. 1, whereas there
is still ambiguity about 2cI. On steric grounds, one would postu-
late that the structure should be as shown. The spectrum for 2b
(Aglycone 1, Fig. 1) shows no signal at 7.55δ, which is evidence for
hydroxyl at C_4 or C_6. It also shows the doublet at 6.7δ for hydroxylation at C_4. The metabolite is thus assigned the structure shown.

(2) ICI 80,996 (CLOPROSTENOL, 'ESTRUMATE')

Cloprostenol, a synthetic prostaglandin analogue (Fig. 2), is
used in cattle in particular for the synchronization of oestrus. It
is a potent agent and the therapeutic dose in cattle is only 500 µg.
Because of the low dose it was decided to investigate the metabolism

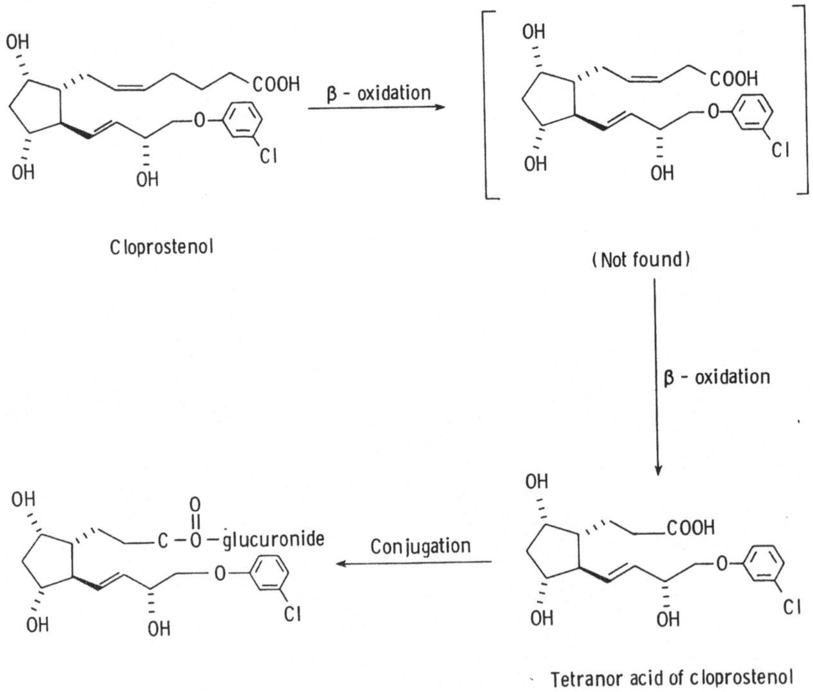

Fig. 2. Cloprostenol metabolism in the cow.

of the drug following administration of 20 times the therapeutic dose
in order to generate a sufficient mass of metabolite for subsequent
identification.

Three cows received a single i.m. injection of 500 µg cloprostenol and 7 days later a 10 mg dose of the drug. Urinary excretion
over 24 h accounted for almost 60% of the dose [1]. The metabolite
patterns in the urine of cows dosed with 0.5 mg or 10 mg was the
same; thus urine obtained following the higher dose was taken for
further analysis.

Isolation of the metabolites of cloprostenol from cow urine
required several chromatographic steps including Amberlite-XAD-2,
Sephadex LH-20 and RP-HPLC [1].

Two major components were isolated, $(B+C)_I$ and $(B+C)_{II}$. GC-MS
was used for identification of the cloprostenol metabolites. During
these studies the isotopic abundance of ^{14}C in cloprostenol has
facilitated metabolite identification.

Low-resolution MS of the TMS derivative of ^{14}C-cloprostenol
showed a prominent ion at m/e 573 $(M-141)^+$ but only a very weak parent ion was detected [2]. Other prominent ions detected were formed by
loss of -O-TMS. The mass spectrum was equivalent to that seen for
authentic cloprostenol except for an increase of two mass units in
each principal fragment detected. The difference in mass of the
fragments between the two isotopic forms was due to the high abundance of ^{14}C in the radiolabelled cloprostenol: the abundance at
position C_{15} was of the order of 90% of the theoretical maximum [2].

.By preparation of mixtures of the $^{12}C/^{14}C$ forms of cloprostenol
characteristic doublets were detected in the mass spectrum of the
dosage solution used in these studies. The presence of these doublets greatly facilitated metabolite identification and removed any
doubt that the metabolites isolated and detected in the mass spectrum
were derived from cloprostenol and not from endogenous prostaglandins. The two major components isolated were identified as cloprostenol $(B+C)_{II}$ and its tetranor acid $(B+C)_I$ (Fig. 2).

During the isolation procedure the major metabolite formed
esters with alcohols used as chromatographic solvents, and under
acidic conditions formed a lactone. All these artefactual products
were treated with base and esterified to form the methyl ester prior
to final confirmation of structure by GC-MS [1].

CONCLUDING COMMENTS

The examples of isolation and identification of Phase I metabolites presented here were chosen to illustrate the use of the perfused liver to isolate Phase I metabolites from bile and studies of the metabolic fate of a veterinary product in which urine was the only reasonable source of metabolites. The examples underline the importance of MS and NMR in such studies. In the metabolism study in the cow the high isotopic abundance of ^{14}C allowed an unusual method of detecting *bona fide* metabolites.

Acknowledgements

The authors gratefully acknowledge the invaluable assistance of the following ICI personnel: David White for radiochemical synthesis, Paul Phillips for mass spectrometry, David Greatbanks and Rod Pickford for NMR analysis, and Colin Lowery for conduct of the *in vitro* studies.

References

1. Bourne, G.R., Moss, S.R., Phillips, P.J. & Shuker, B. (1980) *Biomed. Mass Spec.* 7, 226-230.
2. Bourne, G.R., Moss, S.R., Phillips, P.J., Webster, J.T.A. & White, D.F. (1979) *Biomed. Mass Spec.* 6, 359-360.

#B-2

THE IDENTIFICATION OF AMITRIPTYLINE
AND BENZODIAZEPINE METABOLITES

Ian D. Watson, David R. Jarvie*,
Anne Chambers and Michael J. Stewart

Department of Biochemistry, Royal Infirmary
Glasgow G4 OSF, and
*Department of Clinical Chemistry, Royal Infirmary
Edinburgh EH3 9YW, U.K.

    HPLC peaks in an assay developed for amitriptyline and
its metabolites were of unclear identity,   elucidated by certain
approaches. The accuracy of a semi-quantitative EMIT procedure
for benzodiazepines was validated. Determining the contribution
of metabolites to a drug assay poses problems to a clinically
oriented laboratory, as these studies illustrate.

    The assay of drugs may be complicated by a requirement to iden-
tify and measure active or potentially active metabolites. Problems
associated with drug metabolite assays are not merely technical: it
is often difficult to obtain pure reference materials, and even when
they are available, they may not resolve all the difficulties. We
now give two examples of problems encountered and how we tackled them.

AMITRIPTYLINE

    An HPLC assay for amitriptyline (AMI) and its metabolites had
been developed [1-3] and applied to other tricyclic-structured drugs

in biological fluids [4, 5] including the tricyclic antidepressant imipramine (IMI) and the phenothiazine chlorpromazine (CPZ).

For AMI, IMI and CPZ the metabolism is complex, Phase I consisting of hydroxylation, desmethylation and, to a lesser extent, N-oxidation and dealkylation. The pharmacologically active metabolites of AMI, IMI and CPZ respectively are nortriptyline (NOR; desmethyl-amitriptyline), desmethylimipramine (DMI) and 7-hydroxychlorpromazine (7-OH-CPZ); the most significant metabolites quantitatively are the 10-hydroxy, 2-hydroxy and (some artefactual ?) 5-sulphoxide derivatives respectively. Combinations of the above Phase I reactions occur. Phase II metabolism leads mainly to inactive glucuronide conjugates.

Samples were collected from subjects receiving AMI therapy for depression, but no other drug therapy. The analytical procedure was as previously published [2] with the following modifications: 0.5–1 ml serum or 0.1 ml urine were used, and alkalinization was effected with ammonia (s.g. 0.88); best extraction efficiency was achieved using a lateral shaker (Griffin) for 10 min. All glassware must be freshly silanized before use.

Metabolite characterization

Pure samples of the hydroxy- and desmethyl-metabolites of IMI and CPZ were available [3]; of the AMI metabolites only NOR and 10-hydroxy–AMI were available. All other peaks obtained by HPLC (Fig. 2) were of uncertain identity, and it was to these that various methods of characterization were applied.

Low correlation chromatography

This approach, involving the use of a two-dimensional plot of retention times relative to an internal standard for an HPLC procedure and a GC procedure [6], gave good discrimination between the available metabolites. However, GC-FID traces obtained from patients' samples showed considerable interference.

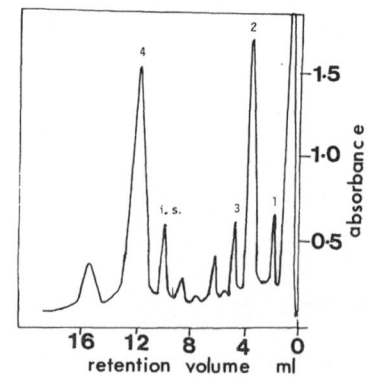

Fig. 1. Typical chromatogram of urinary AMI metabolites. Column: 100 × 5 mm silica (5 µm Hypersil, Shandon; slurry-packed). Other conditions as in ref. [4]. Peak 1, AMI; 2, 7-OH-AMI; 3, NOR; 4, 10-OH-NOR. Identification uncertain for the peaks between NOR and the i.s., viz. protriptyline. (In ref. [4], a correction!)

HPLC followed by MS

Sequential fractions of eluate corresponding to the detected peaks were obtained using a fraction collector. These were dried down under nitrogen and submitted to GC-FID and then to GC-MS in the EI mode.   Both the GC traces and the mass fragmentogram were heavily contaminated with phthalates.

Characteristic HPLC retention changes

The HPLC system developed was an NP system with a silica column and a dichloromethane/propan-2-ol/ammonia (s.g. 0.88) eluent. In this system the retention is in order of increasing polarity (Fig. 1). The expected elution order for the desmethylated metabolites is tertiary-secondary-primary (with reference to the aminopropyl side-chain); but in fact the order was tertiary-primary-secondary for the metabolites of both IMI and CPZ.   The order may be related to the relative contributions of steric and inductive basicity of the terminal amino group.

The influence of systematic changes in eluent on the retention of 10-OH-AMI and NOR was examined.   Fig. 2 shows that the retention of AMI and its metabolites decreased exponentially with increasing solvent polarity and alkalinity; 10-OH-AMI retention decreased more than that of NOR as the proportion of propan-2-ol increased, and NOR retention was similarly affected by increasing concentrations of ammonia, so that there were solvent compositions at which these two compounds were unresolved.   Without the availability of pure samples it would have been impossible to investigate this phenomenon, and it is conceivable that they might have been eluted as a single peak had an arbitrary solvent mixture been chosen.

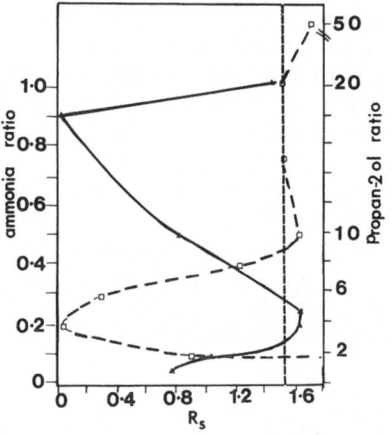

Fig. 2. Effect of changes in the proportion (v/v) of ammonia (—▲—) or propan-2-ol (---□---) on the resolution of 10-OH-AMI and NOR. R_S = resolution function. *From ref. [4], by permission.*

Examination of nortriptyline (NOR) metabolites

A study of the metabolites produced by a patient taking NOR
– a metabolite of AMI – seemed likely to give a clearer   pattern
than that obtained from a patient taking AMI. As there is a report
[7] that 80% of NOR was excreted as 10-OH-NOR, this compound should
constitute the major peak in the chromatogram from a urine sample.

A metabolite of k' = 19.5 was present at high concentration
in urine from patients receiving NOR, and was also present in the
urine of those taking AMI but at lower relative concentrations. Its
concentration after hydrolysis of urine and extraction was raised,
and the ratio of the free metabolite to the conjugate (1:1) was
similar to that reported by Alexanderson & Borgå for 10-OH-NOR. This
identification was assigned to the peak.

A peak corresponding to monodesmethyl-NOR (didesmethyl-AMI)
should also have been present in chromatograms for patients taking
either AMI or NOR. Such a peak with k' = 4.0 was consistently found
although not in blank urine. Its assigned identity is didesmethyl-
AMI, consistent with its retention volume between the tertiary and
secondary amines, AMI and NOR.

The metabolite beyond 10-OH-NOR (Fig. 1; more polar) with k' =
26.0 was not identified; again it showed a concentration increase
following hydrolysis. It might be the dihydroxy metabolite.   The
only other peaks associated with NOR administration were those with
k' = 9.0 and 12.0, probably desmethylated hydroxylated compounds in
view of the metabolic pathways.

What was achievable by the HPLC approach

Endogenous compounds with k's of 10.0 and 14.0 were consistently found.   The identification and quantitation of the AMI metabolites was hampered by the non-availability of authentic samples. Yet
the above use of HPLC gave the following types of guidance.-
1. It is possible to broadly judge compound polarity from the extent
of retention in the chromatographic system.
2. Study of characteristic changes of retention by compounds with
a structure similar to the drug of interest allows inferences to be
drawn as to identity.
3. Study of the metabolic profile following ingestion of a known
metabolite here confirmed the identity of the desmethylated AMI metabolites, and comparison with previously noted patterns allowed
identity to be inferred.

Separation mechanism

The use of chlorohydrocarbon solvents causes quaternization of
the nitrogen in the aminopropyl side-chain following nucleophilic

attack on the chlorohydrocarbon [8].   The mechanism of separation in our HPLC system is probably more complex than it appears and may in fact be a form of 'normal phase ion-pair'. We did not observe any damage to the silica column packing by the basic eluent; damage was probably prevented by the formation of a polar solvent layer on the silica due to hydrogen bonding. This has been discussed previously [4].

BENZODIAZEPINES

Samples were obtained from patients receiving diazepam or temazepam prior to oesophagoscopy, and analyzed by a homogeneous immunoassay and by HPLC. The calibration and use of the enzyme multiplied immunoassay technique (EMIT) was as recommended by Syva, the manufacturer; results are expressed as 'diazepam equivalents'. In this benzodiazepine assay the antibody responds to the general benzodiazepine structure rather than to a specific member of the class; all benzodiazepines are therefore detected, although with differing sensitivities. The response curves shown in Fig. 3 were obtained by preparing standard solutions ranging from 0.3 to 10 µg/ml of the benzodiazepines and metabolites listed in the legend. Since authentic material gave curves not significantly different from that for Syva calibrator, the curve illustrated applies to either diazepam or desmethyldiazepam.

Fig. 4 shows the HPLC approach ⌊9⌋ to measuring diazepam, desmethyldiazepam, temazepam and oxazepam. Scheme 1 shows how some of the benzodiazepines are related to one another.

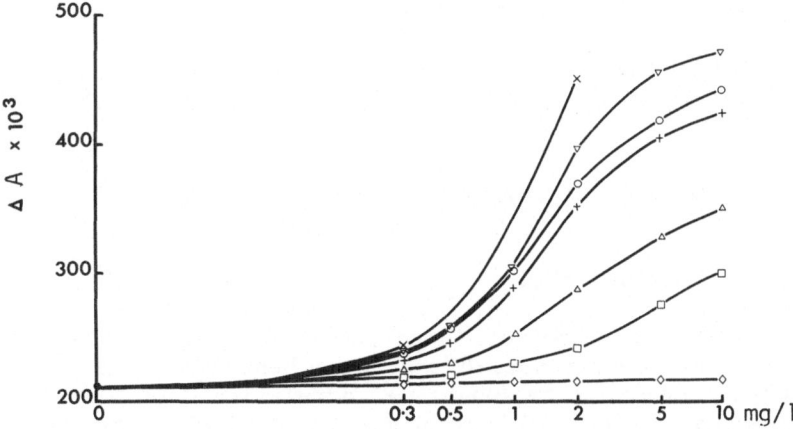

Fig. 3. Cross-reactivities of diazepam and nitrazepam metabolites in serum to the EMIT benzodiazepine antisera.
× kit calibrator, diazepam,   + oxazepam   □ 7-aminonitrazepam
  or desmethyldiazepam   Δ desmethyloxazepam ◇ 7-acetamidonitra-
▽ temazepam   o nitrazepam   zepam.

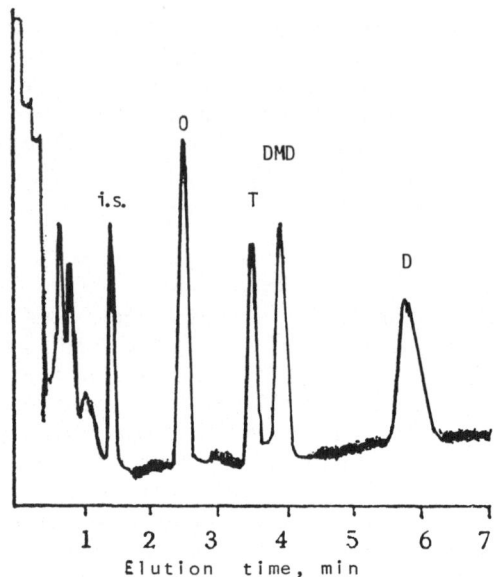

Fig. 4. HPLC of benzodiazepines extracted from a plasma sample into which each compound had been spiked (50 ng|ml).
Column: 10 cm x 5 mm i.d., 5 μm ODS Hypersil. Eluent: methanol/10 mM Na acetate buffer pH 4.6, 6:4 by vol.; 1.7 ml/min. Detection: 234 nm.
i.s. = internal standard (carbamazepine), O = oxazepam, T = temazepam, DMD = desmethyldiazepam, D = diazepam.

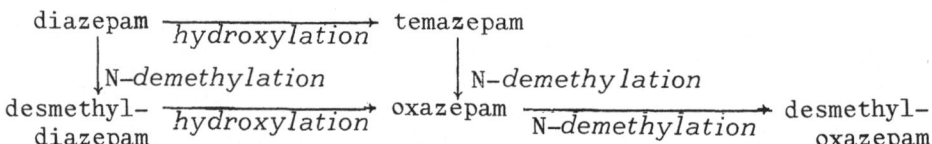

Scheme 1. Some benzodiazepine inter-relationships.

Comparison of EMIT and HPLC

The results in terms of 'diazepam equivalents' obtained using the EMIT kit were compared with the total benzodiazepine (the sum of the 4 species) measured by HPLC. A correlation coefficient of 0.97 was obtained (y = 0.95, x = 0.8; n = 44). A slightly better correlation (0.98) was obtained when the HPLC results were given a benzodiazepine equivalent value from the appropriate standard curve prior to summation and comparison with the EMIT result. In general, the latter are higher than the HPLC values, perhaps due to the contribution of minor metabolites, e.g. desmethyloxazepam, not measured by the HPLC technique.

The good comparability found in the benzodiazepine studies between EMIT and HPLC procedures indicates that, where the identity of the ingested benzodiazepine is known, the value obtained using EMIT for 'total benzodiazepines' will approximate to the sum of unchanged drug and metabolites. EMIT can therefore be used in pilot pharmacokinetic studies provided that the relationship of the metabolites with the time course is appreciated, or as a rapid non-discriminatory screening procedure in cases of overdose.

CONCLUSION

There is no substitute for pure samples of metabolites for use as reference materials, enabling the contributions to measured drug concentrations to be assessed and the identity of metabolites in an assay system to be confirmed. However, if reference material is lacking it is possible to identify metabolites by inference, as we have attempted to do. If our attempts to apply GC and GC-MS had succeeded, we would have had more confidence in the identity of our AMI metabolites and might have been able to establish the cause of the slightly higher EMIT values. The more important metabolic routes for many drugs have already been documented, but confirmation of literature reports and the identification of metabolites in different analytical systems is made unnecessarily difficult if reference materials are lacking. If it is difficult to confirm the identity of major metabolites, the chances of identifying minor ones are hardly high.

It is so important to have metabolites available for reference in clinical investigations that, despite obvious practical difficulties, there is a need for some form of centre, national or international, for drug-metabolite reference standards.

Acknowledgements

Tricyclic drugs and metabolites were given by the following: Drs. J.P. Moody, M. Turnbull, R. Sparks and Professor I. Stevenson (Departments of Biochemical Medicine and of Pharmacology & Therapeutics, Ninewells Hospital and Medical School, Dundee); Dr. A.A. Manion (U.S. National Institute of Mental Health, Bethesda, MD); H. Lunbeck and Co. (Copenhagen-Valby, Denmark); Smith, Kline & French Laboratories (Welwyn Garden City, U.K.); and May & Baker Ltd. (Dagenham, U.K.). We also thank Roche Products Ltd. (Welwyn Garden City) for a gift of benzodiazepines and metabolites.

References

1. Watson, I.D. & Stewart, M.J. (1975) *J. Chromatog.* 110, 389-392.
2. Watson, I.D. & Stewart, M.J. (1977) *J. Chromatog.* 132, 155-159.
3. Watson, I.D. & Stewart, M.J. (1977) *J. Chromatog.* 134, 182-186.
4. Watson, I.D. (1979) *Proc. Analyt. Div. Chem. Soc.* 16, 293-297. *(See overleaf for correction.)*

Author's correction of error: in ref. 4 the solvent composition in the legend to Fig. 4 should read *(agreeing with the text)*   Dichloromethane:propan-2-ol:ammonia 100: 1.75:0.2.

5. Moyes, R.B. & Moyes, I.C.A. (1977) *Postgrad. Med. J. 53,* 431-439.
6. Hucker, H.B. & Stauffer, S.C. (1974) *J. Pharm. Sci. 63,* 296-297.
7. Alexanderson, B. & Borgå, O. (1973) *Eur. J. Clin. Pharmacol. 5,* 174-180.
8. Beckett, A.H. (1974)   *The Poisoned Patient: The Role of The Laboratory,* Ciba Found. Symp. No. 26 (Porter, R. & O'Connor, M., eds.), Elsevier, Amsterdam, p. 79 (Discussion following article on pp. 57-76).
9. Douglas, J.G., Nimmo, W.S., Wanless, A.R., Jarvie, D.R., Heading, R.C. & Finlayson, N.D.C. (1980) *Br. J. Anaesth. 52,* 811-815.

#B-3

A NOVEL USE OF RIA AND HPLC IN THE IDENTIFICATION
OF PLASMA METABOLITES OF HR 158

J.D. Robinson, I.D. Wilson, C.D. Bevan and J. Chamberlain

Hoechst Pharmaceutical Research Laboratories
Milton Keynes, Bucks. MK7 7AJ, U.K.

The benzodiazepine HR 158 is active at dose levels of 1 mg, resulting in peak plasma levels of 5-6 ng/ml. Although radio-labelled metabolism studies in animals had indicated the nature of the main metabolites, these could not be confirmed in man due to the obligatory low doses. Moreover, conventional modes of analysis such as HPLC or GC were too insensitive for metabolite detection and identification in human plasma. The high sensitivity of radioimmunoassay (RIA) was used to overcome this problem. Of the two antisera available, one was highly specific for HR 158 furnishing plasma levels similar to those found by a specific HPLC method, and one was non-specific yielding much higher values and thus indicating the presence of cross-reacting metabolites. The latter was used on eluate fractions in an HPLC system capable of separating the known and putative metabolites in human plasma. The resulting RIA-chromatogram allowed identification at ng/ml level.

HR 158 is a benzodiazepine analogue whose major use is as a hypnotic, although it has been used intravenously for the treatment of post-operative pain. The chemical structures of HR 158 and its known and putative metabolites are shown overleaf. The low doses employed meant that a new system for the detection of metabolites in human plasma had to be developed. This was achieved by combining the high sensitivity of a RIA method with the selectivity of HPLC.

RIA OF HR 158

Two antisera have been raised against HR 158 immunogens. The specific antiserum was raised against a conjugate of HR 158 to bovine serum albumin through the 3-position of the benzodiazepine ring, while the non-specific antiserum was raised against a conjugate

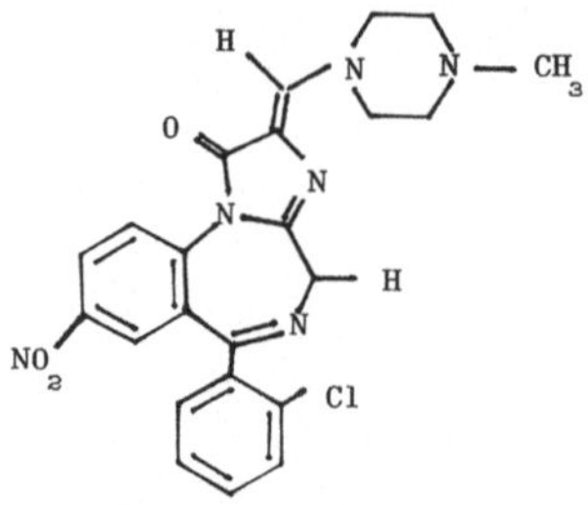

HR 158

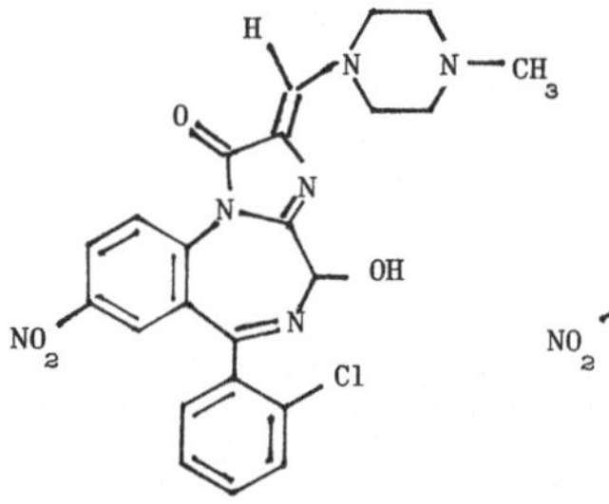

RU 1633

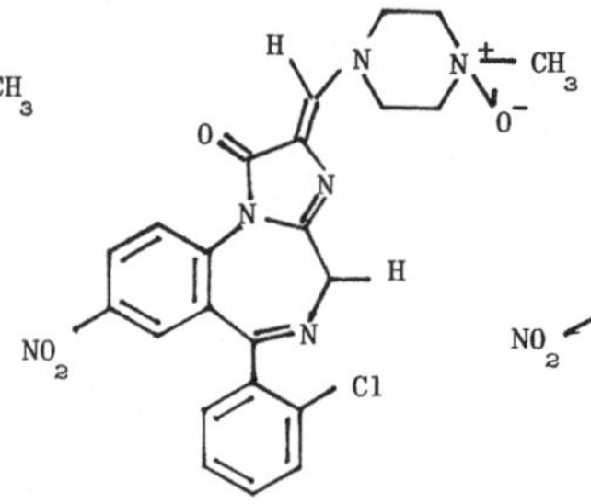

CRO 7966C
(N-OXIDE)

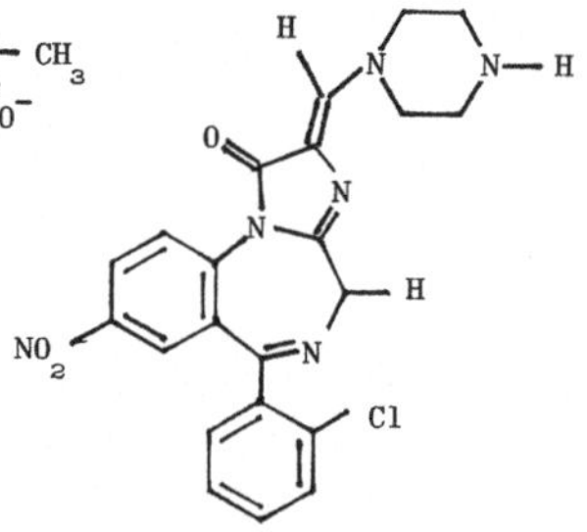

RU 1490

through the methyl group. The latter was used in this study since it cross-reacted with all of the putative metabolites (*vs.* HR 158 as 100%: *N*-oxide, 156.5%; RU 1490, 75.0%; RU 1633, 1163.8%). The RIA was based on previously described procedures [1] and was performed in 0.1 M phosphate buffer containing 0.5% bovine serum albumin. Tritiated HR 158 was used as the tracer (s.a. 9.0 Ci/mmol; Hoechst AG) and the assay standards were prepared in buffer, from an ethanolic stock solution, to cover the range 0.5 ng/ml to 250 ng/ml. The non-specific antiserum (No. 47-12/7/78) was used at an initial dilution of 1:55. A 100 μl aliquot of samples was used, and after an incubation period of 4 h, phase separation was achieved by the addition of dextran-coated charcoal followed by centrifugation.

HPLC OF HR 158

The basis of the chosen system was a 10 cm X 2 mm i.d. column packed with 5 μm ODS-Spherisorb and a solvent flow rate of 0.8–1.0 ml/min. The eluate from the column was collected at ½-min intervals (~0.5 ml) into small tubes and was later subjected to the RIA.

DEVELOPMENT AND CAPABILITIES OF THE METHODS

Choice of solvent

Since organic solvents are known to affect the antigen–antibody interaction, it was essential to establish which solvent system

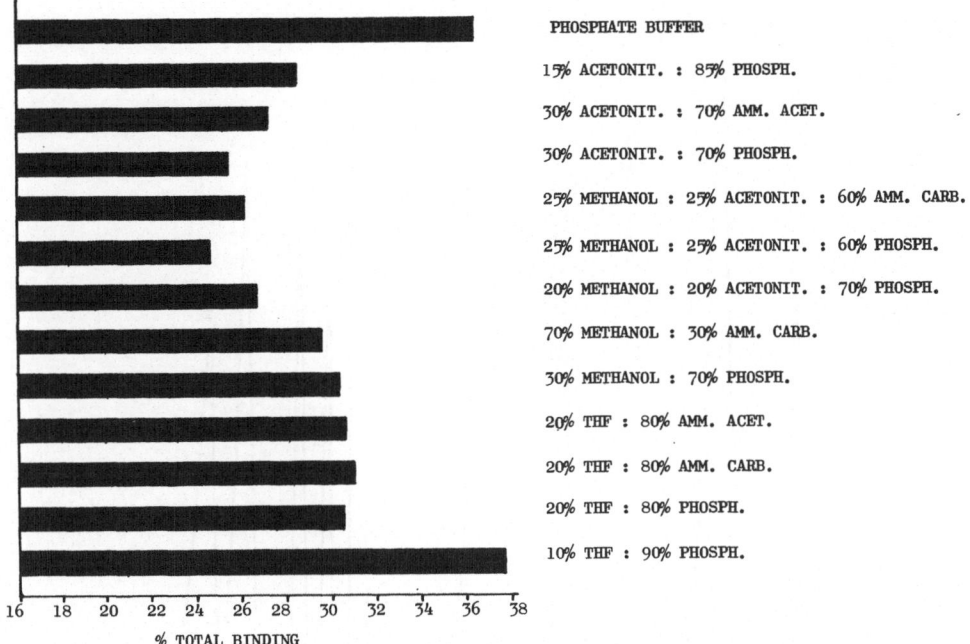

Fig. 1. HPLC solvent systems and their effect upon the binding of HR 158 to antibody at zero concentration. THF denotes tetrahydrofuran; the aqueous component consisted of 0.1 M phosphate or of ammonium carbonate solution (pH of former = 7.4; solvent aliquots = 100 μl).

would suit best for both RIA and HPLC. Accordingly, the degree of binding of the non-specific antibody to the tracer was assessed in the presence of aliquots of many different solvents, as shown in Fig. 1.

From this experiment it was concluded that three systems were worthy of further investigation, viz. 10% THF : 90% phosphate, 15% acetonitrile : 85% phosphate, and 30% methanol : 70% phosphate.

Optimization of HPLC conditions

With the chosen solvent systems, the HPLC separation of HR 158 and its putative metabolites was then studied. Minor modifications were made to the solvent ratios in order to achieve optimum separation. For each system as thus modified, Fig. 2 shows the resulting UV chromatogram. The results indicated clearly that the most useful chromatographic separation was achieved with 20% THF : 80% phosphate, as adopted for the rest of the study.

Preparation and analysis of samples

Since the low dosage used for HR 158 produces low levels of the

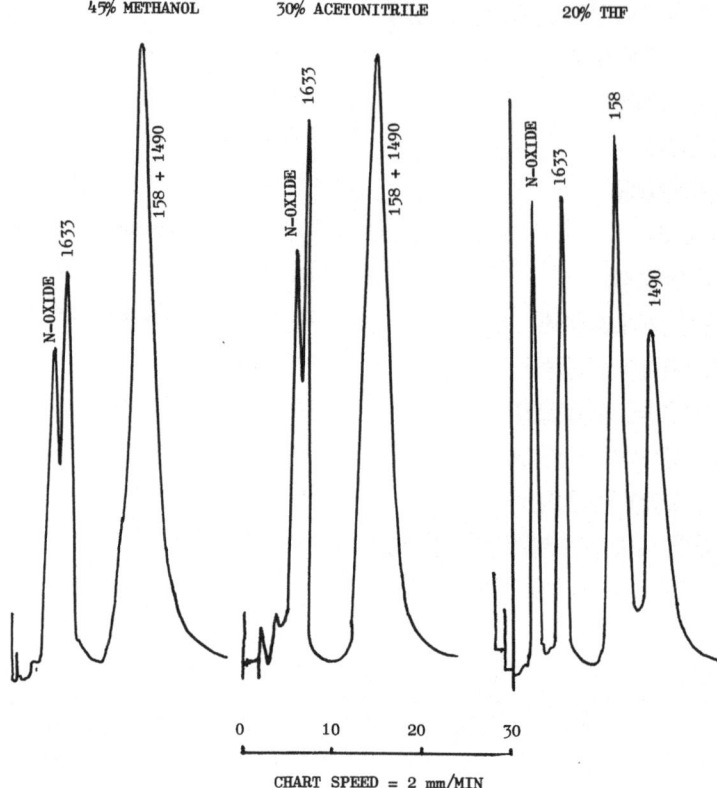

Fig. 2. HPLC-UV patterns with different solvent systems for elution. For conditions, see text.

parent compound, and could not be increased to increase the levels of possible metabolites, plasma samples collected during a multi-dose clinical trial were pooled and extracted. The subjects received 2 mg once a day for 3 months, and blood samples were collected twice a day. Aliquots of the samples from the middle of the study were mixed to provide a 70 ml pooled sample (**S**).

In order to ascertain that the putative metabolites, if present in the sample, would be detected in the final analysis, a pool (**P**) of normal plasma to which known amounts of the putative metabolites were added was prepared (10 ng/ml).

Aliquots (35 ml) of both samples (**S** & **P**) were then subjected to the same extraction procedure. Firstly nitrazepam was added to each aliquot to enable the extraction and chromatographic separations to be studied, i.e. to act as an i.s. (nitrazepam shows minimal RIA cross-reactivity). To each sample was added 350 ml of ethyl acetate,

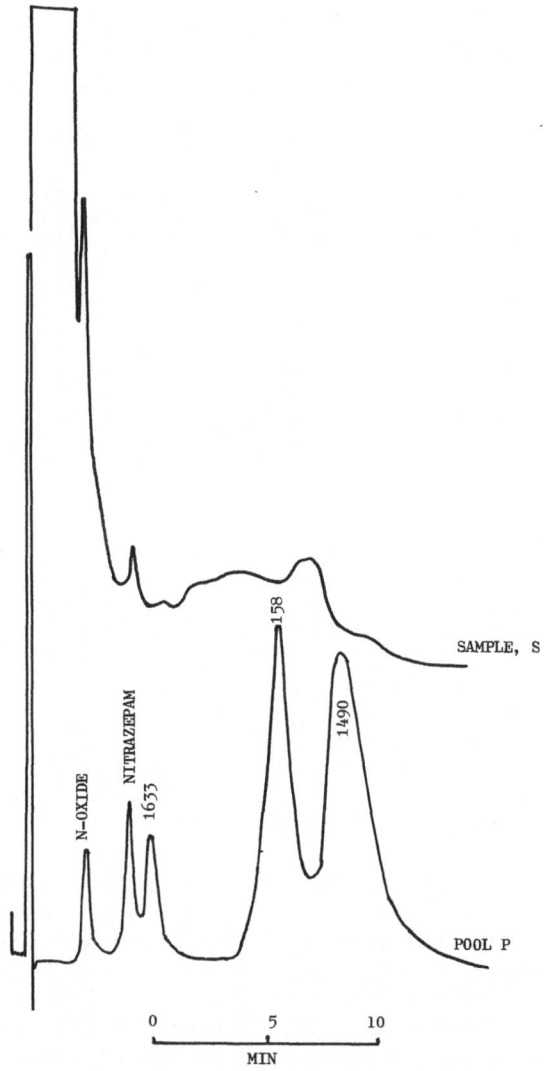

Fig. 3. HPLC-UV patterns with the 20% THF:80% phosphate eluent, for plasma from the trial (**S**) or with spiked-in compounds (**P**).

which had been shown to quantitatively extract each of the putative metabolites from plasma. The samples were then centrifuged and the aqueous layers discarded. The organic layers were evaporated to dryness, initially by rotary evaporation to reduce the volume and finally under nitrogen. The residue was resuspended in 1.0 ml of methanol, and 25 µl aliquots were injected onto the HPLC column. The fractions collected over 30 min were assayed in duplicate by the non-specific RIA procedure.

The UV patterns for the 2 samples as they were eluted from the column (Fig. 3) show very good separation for the spiked sample **P**. The pattern for the pooled sample **S** was difficult to interpret. The

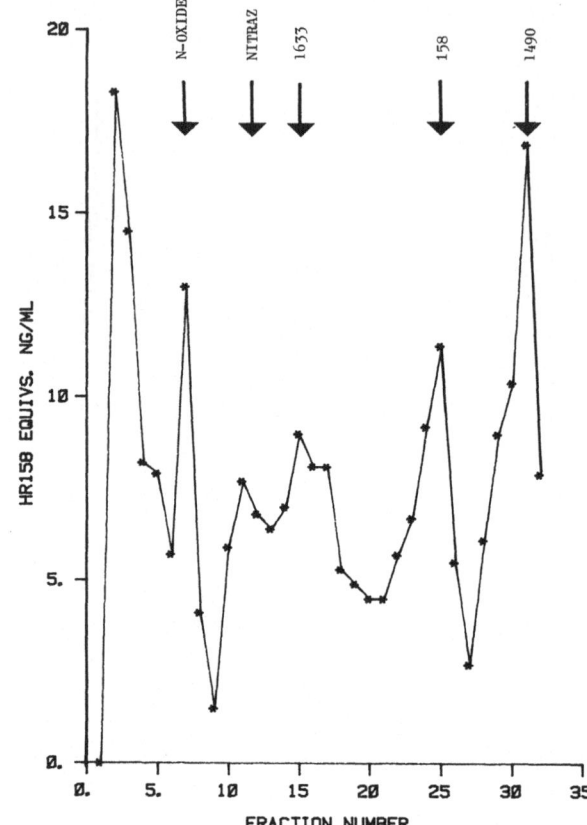

Fig. 4A. HPLC–RIA pattern
for the spiked plasma **P**,
corresponding to the lower
UV trace in Fig. 3.

low levels of HR 158 and any of the metabolites present did not
yield a clear UV trace. However, the i.s. could be detected, indicating that both the extraction and chromatography had been successful.

    Figs. 4A & 4B illustrate the results of the RIA analysis of the
individual fractions from the eluates, plotted as concentration *vs.*
fraction no. ('HPLC–RIA–O-gram').   Concentrations are plotted as
equivalents of HR 158 because pure HR 158 was used as the standard
and the putative metabolites have different cross-reactivities with
respect to HR 158.

    For **P** the RIA pattern (Fig. 4A) showed that each of the com-
pounds had been eluted from the HPLC column and could be detected by
RIA, and there was a close resemblance to the UV pattern. The large
peak seen in fractions 1 and 2 was due to the solvent front effect –
the combined effect of the mobile phase and the methanol injected
with the sample. The extra solvent reduced the binding of the antigen to the antibody, hence giving a falsely high value for concentration.   The trace for the sample **S** demonstrated a lower solvent

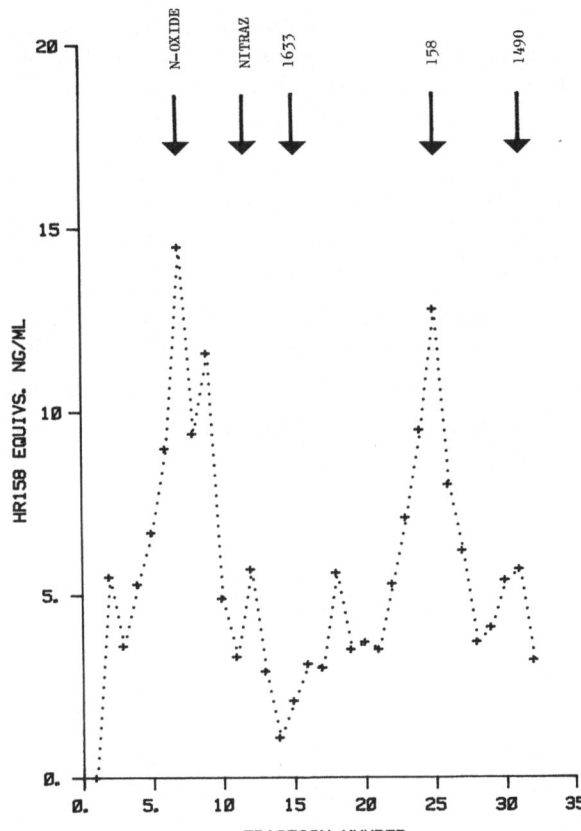

Fig. 4B. HPLC–RIA pattern for the pooled plasma S, from the clinical trial, as in Fig. 3 where the upper trace shows the UV pattern. Marker positions also shown.

front effect. The peaks for the *N*-oxide, HR 158 and RU 1490 could clearly be distinguished, but no peak could be seen corresponding to RU 1633.

DISCUSSION

The HPLC–RIA patterns show that HPLC eluates can be subjected to a sensitive RIA procedure without having to remove the solvent, e.g. by freeze-drying, and reconstituting. This can only be done, however, if the solvent system is chosen carefully and its effect on the RIA fully evaluated prior to its adoption. Here the little-used organic modifier THF was the best choice, providing the best HPLC separation and having the least effect in the RIA.

The study has also demonstrated that the metabolites of HR 158 found in man are the *N*-oxide and RU 1490. Since the *N*-oxide has a cross-reactivity of 156.5% using this antiserum with respect to HR 158, the concentration of *N*-oxide seen in the HPLC–RIA pattern is a slight overestimation and the true value is about two-thirds of that shown. Similarly RU 1490 is grossly overestimated since its cross-reactivity is 1164%.

The approach of HPLC followed by RIA has also been successfully used to study the presence of HR 158 and its metabolites in samples of bile collected after post-operative administration of HR 158 (J.D. Robinson & H.P.A. Illing, unpublished data).

The HPLC-RIA approach has, then, revealed HR 158 metabolites in man, viz. *N*-oxide and (minor) RU 1490 but no hydroxy compound RU 1633.

References

1. Robinson, J.D. (1975) *The Development of Radioimmunoassay for Therapeutic Drugs*, Ph.D. Thesis, University of Surrey, U.K.

Additional reference to the RIA approach:

2. Robinson, J.D., Morris, B.A., Aherne, G.W. & Marks, V. (1975) *Br. J. Clin. Pharmacol. 2*, 345-349.

#B-4

INVESTIGATION OF PESTICIDE METABOLITES IN THE HUMAN

C.V. Eadsforth

Shell Research Ltd.
Sittingbourne Research Centre
Sittingbourne, Kent ME9 8AG, U.K.

In pesticide manufacture and usage, human exposure is minimized at every stage. The need to monitor this is the main impetus to the study of pesticide metabolism in man. That of selected pesticides, initially in animals, has been studied as part of industrial hygiene research programmes to indicate the most suitable analytical methods for monitoring exposure. These typically entail GC-ECD, and their development is in general based on a common major urinary metabolite or on the pesticide itself. Dose-excretion studies enable assays to be made on manufacturing operatives, formulators and sprayers.

Naturally occurring and synthetic pesticides are very diverse in chemical type. They broadly fall into the following groups:- insecticides, nematocides, fungicides, rodenticides and herbicides. The major classes of insecticides include the organophosphates (e.g. parathion, diazinon), the carbamates (e.g. carbaryl), the organo-chlorines (e.g. DDT, dieldrin) and now the pyrethroids (e.g. per-methrin, cypermethrin). Other groups include natural toxin deriva-tives (e.g. rotenone, nicotine) and the growth regulants (e.g. the juvenile hormone analogues).

Extrapolation of pesticide toxicological data to man is, at present, by judging the outcome of various toxicity and metabolism studies in several animal species. If there are no observed prob-lems and if there is a uniformity of metabolism in several species, a compound is conventionally assessed as low risk. In contrast, the development of a drug necessitates some human experimentation between animal experimentation and clinical trials. Pesticide development is different in that (i) human exposure is minimized at all stages of use, and (ii) there is nothing equivalent to a therapeutic dose at which metabolism/toxicity studies should be done. The main impetus for investigation of the metabolism of pesticides in man stems from the desirability of biological monitoring. This involves the measurement of *internal* exposure through the analysis of the primary agent or its metabolites in a biological specimen such as urine, faeces, blood or expired air.

Several examples are given of the metabolism of selected pesticides, initially in animals and subsequently in man, incorporating dose-excretion studies wherever feasible. In general, a common major urinary metabolite or the pesticide itself has then been used as the basis for developing an analytical method which is subsequently used quantitatively for monitoring exposure amongst employees handling the pesticide, assuming that human dose-excretion data are available. Such studies not only provide what is sought, i.e. a relationship between exposure and analytical values, but also afford confirmation of metabolic pathways discovered in animals.

ORGANOPHOSPHORUS INSECTICIDES

These insecticides are neutral esters of phosphoric acid or its thio analogues. Many animal studies have shown the anticipated rapid metabolism, and some work has been done in man. Urinary elimination of radioactivity was 40-90% in 5 days after i.v. administration of pesticides including monocrotophos, ethion, azinphosmethyl, malathion and parathion [1]; thus the metabolic capacity is high in man too.

Monocrotophos

Monocrotophos* and related insecticides such as dichlorvos and mevinphos form in animals a common urinary metabolite, dimethyl phosphate. This was the main individual metabolite during 2 days in rats given [^{32}P]monocrotophos, its yield exceeding 4-fold that of O-desmethyl monocrotophos [2]. The main metabolic reactions are hydrolysis of P-O-C linkages to form dimethyl phosphate and O-desmethyl monocrotophos (Fig. 1) [3]. Desmethylation of monocrotophos via its N-hydroxymethyl derivative affords the unsubstituted amide analogue N-desmethyl monocrotophos. It has been proposed that dimethyl phosphate excretion be used as a measure of human occupational exposure to monocrotophos.

Previous methods for determining dimethyl phosphate in urine have relied on alkylation of the metabolite to yield a volatile trialkyl phosphate [4, 5]. A difficulty with urine samples is that they are naturally rich in inorganic phosphate. Unless this is removed, treatment with diazomethane results in so much trimethyl phos-

---

* AZODRIN, a Shell Registered Trade Mark; others:- chlorfenvinphos - SUPONA, BIRLANE; flamprop-methyl - MATAVEN; flamprop-isopropyl - BARNON, and its R(-) isomer - SUFFIX BW; benzoylprop-ethyl - SUFFIX.

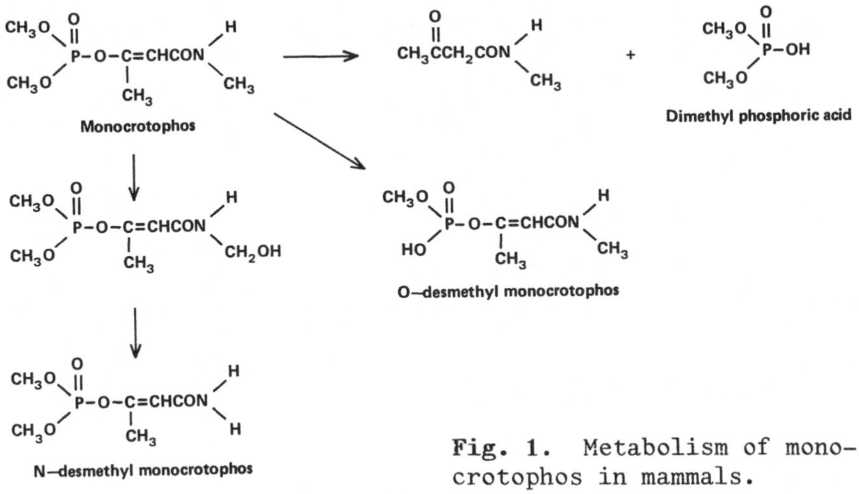

Fig. 1. Metabolism of mono-crotophos in mammals.

phate that final chromatographic peaks corresponding to the derivatized metabolites may be masked. Earlier methods have employed solvent extraction to diminish this interference; but the partition coefficients are at best unfavourable to the extraction of dimethyl phosphate and moreover the procedures are inefficient in excluding the inorganic phosphate. This has been overcome by using diazopentane to yield derivatives that are more easily resolved by chromatography [4]. However, the higher diazoalkanes are relatively inaccessible and costly. The following method [6] is less time-consuming and is more accurate than hitherto published methods.-

(1) Urine treated with calcium hydroxide and centrifuged, overcoming interference from endogenous phosphate by partial selective removal.

(2) Supernatant acidified with cation-exchange resin.

(3) Ethylate directly with diazoethane (the use of ion-exchange resin obviates the risk of incomplete derivatization of the dimethyl phosphate that is inherent in the earlier methods).

(4) Apply GC–NPD to determine the ethyldimethyl phosphate, which is well separated from triethyl phosphate formed by ethylation of any residual inorganic phosphate; with an N–P bead detector operating in the phosphorus mode, rather than a flame photometric detector (FPD), 0.01 µg/ml urine is now detectable.

Chlorfenvinphos

For this member of the vinyl phosphate group of insecticides, a specific chemical method to monitor human exposure has recently been developed at Shell Toxicology Laboratory; previously depression of blood cholinesterase or electroneural physiological measurements were used as indicators of exposure [7].

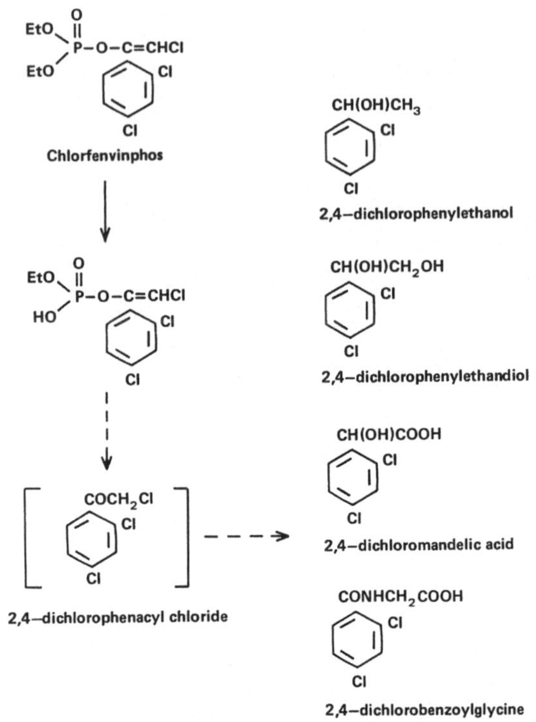

Fig. 2. The metabolism of chlorfenvinphos in the mammal.

Metabolism studies in rats and dogs [8] and *in vitro* studies with liver enzymes [9] have revealed the major transformation pathways (Fig. 2). The primary detoxification reaction was oxidative desethylation to afford desethylchlorfenvinphos which was partly excreted and partly further metabolized via 2,4-dichlorophenacyl chloride and 2.4-dichloroacetophenone to 2,4-dichlorophenylethanol, the ethandiol, 2,4-dichloromandelic acid and 2,4-dichlorobenzoylglycine. The excretion of [^{14}C]chlorfenvinphos was then studied in a male subject. An oral dose of 0.18 mg/kg was almost totally eliminated (72% in 4.5 h; 94% in 26.5 h). The yields of the urinary metabolites are compared with those in the rat, dog and rabbit in Table 1. Each species had its distinctive metabolite profile, maybe partly reflecting dose-dependency.

At this stage desethylchlorfenvinphos, 2,4-dichloromandelic acid and 2,4-dichlorophenylethanol glucuronide were candidates for use in monitoring occupational exposure to chlorfenvinphos. The first was precluded when its excretion was found to average only 5% of a non-radioactive chlorfenvinphos dose in a further study with 14 male subjects, performed to relate the yield of a specific metabolite to amount ingested [10]. An alternative approach to the problem involved conversion of the mixture of urinary metabolites into a single degradation product 2,4-dichlorobenzoic acid using alkaline potassium permanganate. Urine from rats dosed with [*phenyl*-^{14}C]chlorfen-

Table 1. Comparative metabolism of [^{14}C]chlorfenvinos: excretion products as % of dose. *From ref. [8].*

Agent *(first entry)* or metabolite	Species (& mg/kg administered)			
	Rat (2.0)	Dog (0.3)	Rabbit (4.0)	Man (0.18)
Chlorfenvinphos	0	0	0	0
Desethylchlorfenvinphos	27*	65*	47	22*
2,4-Dichloromandelic acid	5	12	3	22*
2,4-Dichlorophenylethanol†	30	3	28	21
2,4-Dichlorophenylethandiol†	2	2	3	14
2,4-Dichlorobenzoylglycine	3	<1	1	4

* Isotope dilution analysis on whole urine.
† As glucuronide conjugates.

vinphos of known specific radioactivity was used to validate the oxidation procedures, monitoring the conversion by GC-ECD. Work-up was typically as follows.-
(1) Treat the urine with alkaline permanganate at 50° for 18 h.
(2) Acidify and extract 3 times with diethyl ether.
(3) Dry down the organic phase and methylate: sulphuric acid/methanol under reflux for 1½ h (gives methyl ester in quantitative yield).
(4) Dilute with water and extract with hexane.
(5) Analyze organic phase by GC-ECD; detection limit ~0.01 µg/ml.

This method is now used for the indication of human exposure to chlorfenvinphos; but other foreign compounds are potential precursors of 2,4-dichlorobenzoic acid. In particular, 2,4-dichlorobenzyl alcohol is an antiseptic component (1 mg) of some throat tablets. We have no data on the metabolism of this compound by man, but it is likely to be excreted as the acid or compounds oxidizable to it. A single tablet might therefore give rise to ~0.5 µg/ml of the acid in urine, as must be kept in mind if the acid is found.

PYRETHROIDS

The synthetic pyrethroid insecticides have been developed from the natural pyrethrins by structural modifications aimed at retaining bioactivity and increasing photostability. Deltamethrin, fenvalerate, permethrin and cypermethrin represent a new generation in commercial use, including veterinary (Fig. 3). Their metabolic fate has been studied in rat, mouse, dog, goat and cow [11-13] and a clear, if complex, picture is emerging. The pyrethroids are, not surprisingly, metabolized by ester-cleavage to their carboxylic acid and alcohol moieties. With cyano-pyrethroids as for the three in Fig. 3 the alcohol is unstable and is rapidly converted into an aldehyde, which is largely oxidized *in vivo* to 3-phenoxybenzoic acid. There

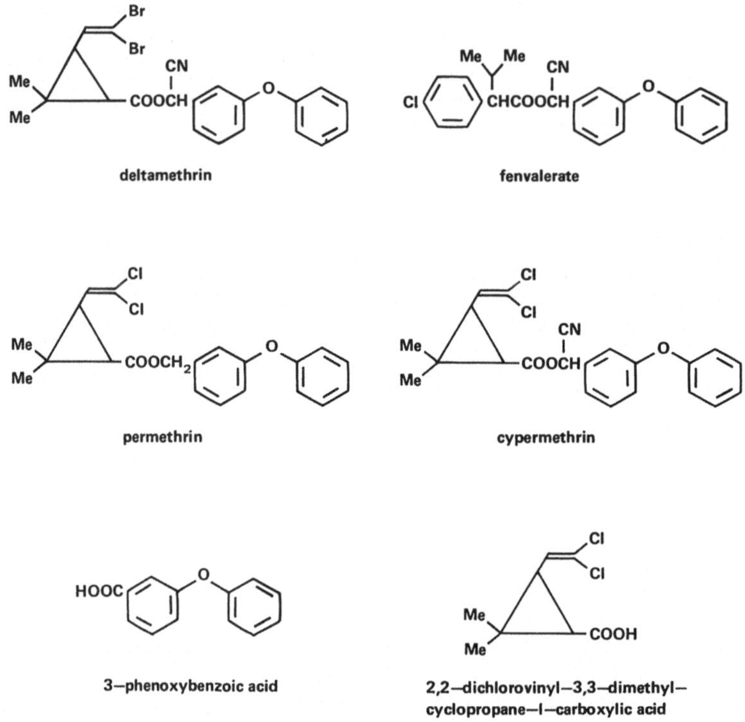

Fig. 3. Structures of some pyrethroid insecticides and associated metabolites.

can be pyrethroid hydroxylation before and after ester cleavage, but apparently as a major reaction only in rats (and to a lesser extent in mice).

Cypermethrin

The animal data helped considerably in selecting a suitable metabolite for monitoring human exposure to cypermethrin. In the 4 species studied in detail to date, the cyclopropane-COOH moieties are excreted (as glucuronides) in the urine in 50-70% yields, and they were the clear choice for a monitoring method. Moreover, the 3-phenoxybenzyl moiety was clearly unsuitable without further study. 3-Phenoxybenzoic acid exhibits a remarkable species specificity in its conjugation, the major metabolite being 3-(4'-hydroxyphenoxy)-benzoic acid sulphate in the rat [14], 3-phenoxybenzoyl taurine in the mouse [15], 3-phenoxybenzoyl glycine in the dog [12], and 3-phenoxybenzoyl glutamic acid in the cow [16]. This array of metabolites renders the conjugation of 3-phenoxybenzoic acid in man unpredictable.

Metabolite yield *vs.* pyrethroid intake has been studied in this laboratory. Male subjects received single oral doses (0.25–1.5 mg) of a 1:1 mixture of *cis* and *trans* cypermethrin, and urine was monitored for free and conjugated carboxylic acid as formed metabolically (named in legend to Fig. 4). The method was as follows.-
(1) Urine dried down and methylated: sulphuric acid/methanol, reflux $1\frac{1}{2}$ h.
(2) Dilute with water, and extract 3 times with hexane.
(3) Chromatograph the organic phase on activated silica with acetone/hexane.
(4) Analyze the eluate by GC–ECD (Fig. 4).

Urinary excretion of the individual *cis* and *trans* isomeric metabolites averaged 78% and 49% respectively in 24 h, near the values

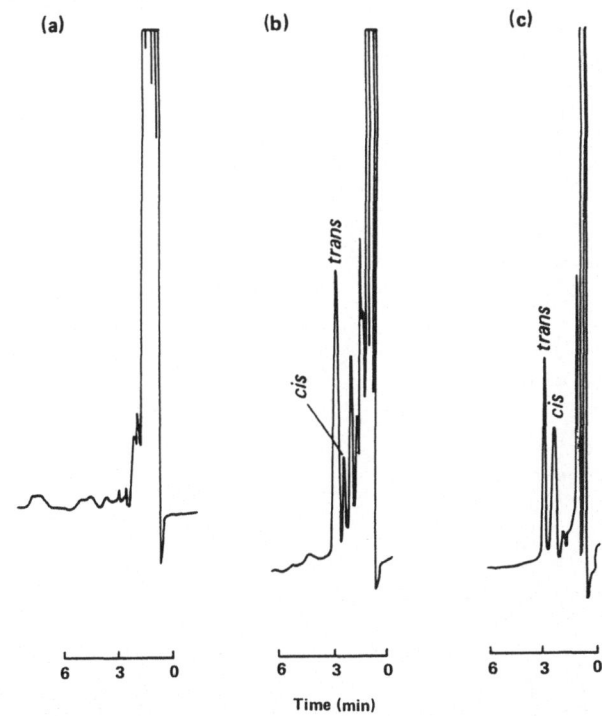

Fig. 4. Typical GC patterns for the methyl esters of 2,2-dichlorovinyl-3,3-dimethylcyclopropane-1-carboxylic acid isomers.
(a) Extract from pre-exposure human urine after methylation.
(b) Urine extract, after methylation, from subject given a 1.5 mg oral dose of cypermethrin.
(c) Isomer standards, *cis* and *trans*: 0.26 ng each (methyl esters). GC conditions: 2 m × 4 mm i.d. column of 4% GE XE-60 on DiatoportS, 80–100 mesh; 120°; 100 ml nitrogen/min.

obtained in mice, and was similar for the different dosages. These values confirm that ester cleavage is a major route of metabolism and that, as in animals, the *trans* isomer is more labile than the *cis* isomer. The amount of the free and conjugated forms of the *trans* cyclopropane-carboxylic acid can therefore serve as a roughly quantitative index of ingestion of cypermethrin in the preceding 24 h.

WILD OAT HERBICIDES

These herbicides are shown in Fig. 5. The elimination of radioactivity is rapid in rats, dogs, mice and rabbits given single oral doses of [*phenyl*-^{14}C]flamprop-methyl (2.5-4.4 mg/kg body wt.) [17]. The structures of the metabolites indicate that the efficient metabolism is due largely to the hydrolysis of the herbicide to the readily excretable carboxylic acid (Fig. 5). This acid is also eliminated partly as a glucuronide conjugate. The reaction dominates the metabolism of the herbicide in each of the mammals studied. Aromatic hydroxylation at the *N*-benzoyl group of flamprop-methyl

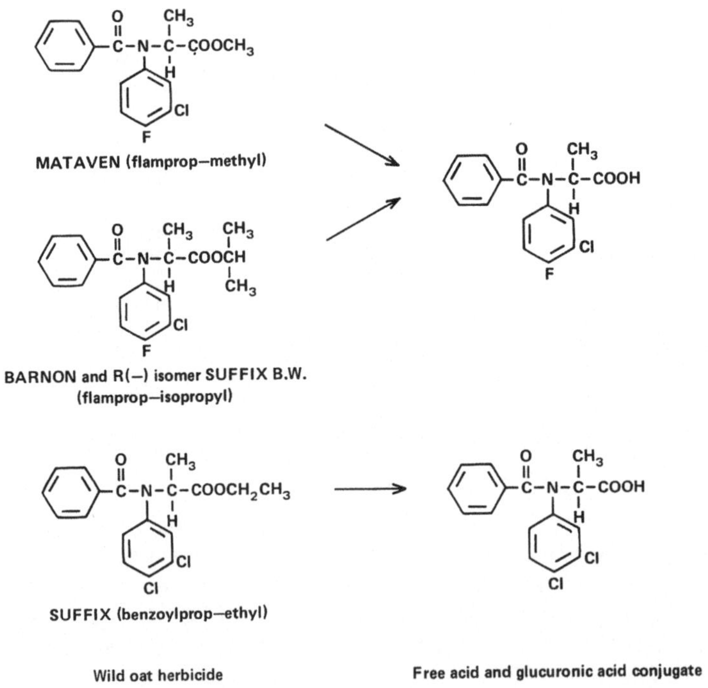

MATAVEN (flamprop—methyl)

BARNON and R(—) isomer SUFFIX B.W.
(flamprop—isopropyl)

SUFFIX (benzoylprop—ethyl)

Wild oat herbicide

Free acid and glucuronic acid conjugate

Fig. 5. Structures of some wild oat herbicides and associated common urinary metabolites.

Table 2. Comparative metabolism of [^{14}C]flamprop-methyl: excretion products as % of dose, viz. 2.5-4.4 mg/kg. *From ref. [17].*

% of administered radio-activity found after 4 days	Species: males and *(italicized)* females			
	Rat	Dog	Mouse	Rabbit
In urine	18	43	30	
	33	*17*	*23*	*81*
In faeces	79	48	60	
	62	*76*	*69*	*13*
As unchanged flamprop-methyl	4	41	7	
	3	*61*	*7*	*4*
As hydrolysis product	24	28	28	
	29	*15*	*17*	*32*
As conjugates	34	13	45	
	42	*15*	*56*	*39*
As non-conjugated hydroxylation products	19	7	6	
	20	*5*	*4*	*1*

also occurs, affording 3-hydroxy, 4-hydroxy and 3,4-dihydroxybenzoyl derivatives. A comparative summary of urinary elimination, faecal elimination, hydrolysis and aromatic hydroxylation collates the data (Table 2).

A striking species difference observed was the relatively large amount of aromatic hydroxylation that occurred in the rat. Another species difference, the relatively high proportion of unchanged flamprop-methyl excreted in dog faeces, was probably the result of limited absorption of the compound. The mouse was rather similar to the dog and to the female rat in its urinary and faecal excretion of radioactivity. The rabbit effected very little aromatic hydroxylation and was unique in that the major route of elimination was the urine. This is, however, a relatively common phenomenon in this species, and is related to a rather high mol. wt. threshold (475±50) for biliary excretion of foreign compounds in comparison with, e.g., the rat (325±50) [18]. In this respect man resembles the rabbit rather than the rat. Hence it would be reasonable to suppose that man would excrete wild oat herbicides, e.g. flamprop-methyl, mostly as a mixture of the free acid and β-glucuronide in urine.

Analytical methods for monitoring human exposure to wild oat herbicides, flamprop-methyl, flamprop-isopropyl and benzoylprop-ethyl, have been based on the determination of the respective free and conjugated urinary metabolites, *N*-benzoyl-*N*-(3-chloro-4-fluorophenyl)-

2-aminopropionic acid and *N*-benzoyl-*N*-(3,4-dichlorophenyl)-2-amino-
propionic acid.   The methods are as follows.-
(1) Urine dried down and methylated: sulphuric acid/methanol, reflux
1 h (converts both the free acid and its ester-type glucuronide into
the corresponding methyl ester).
(2) Dilute with water, and extract twice with diethyl ether.
(3) Dry down the organic phase, and dissolve residue in hexane.
(4) Chromatograph on deactivated Florisil with acetone/hexane.
(5) Analyze the eluate by GC-ECD (Fig. 6; detection limit well below
0.01 µg/ml urine in both cases, and control urine does not give peaks
corresponding to the acids).

3-Chloro-4-fluoroaniline (CFA)

CFA is an intermediate in the manufacture of the two wild oat

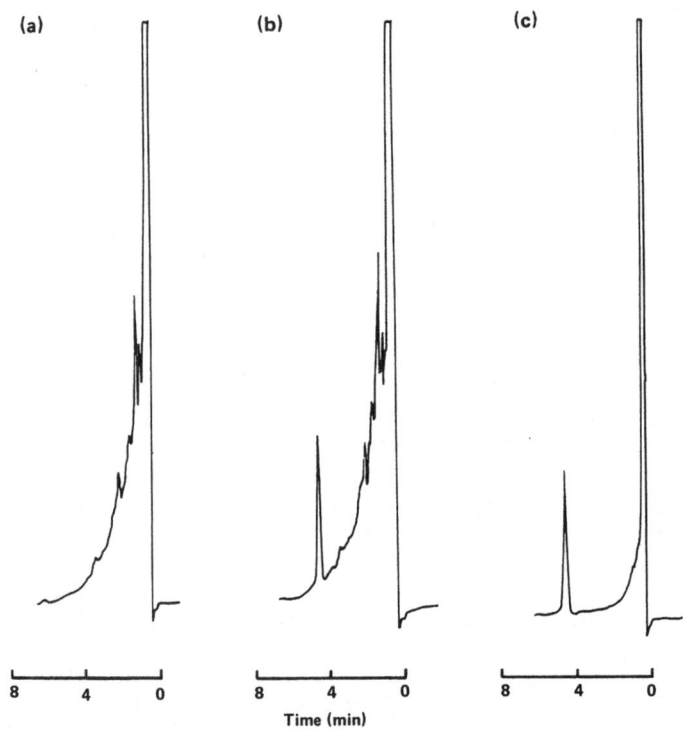

Fig. 6. Typical GC patterns for the methyl ester of *N*-benzoyl-*N*-(3,4-
dichlorophenyl)-2-aminopropionic acid.
(a) Extract from control rat urine after methylation.
(b) Urine extract, after methylation, from rat dosed with   benzoylprop-ethyl.
(c) Standard, as methyl ester in hexane: 0.11 ng.
GC: 2 m × 4 mm col., 3% OV-225 on Gas Chrom Q, 80-100; 250°; N_2 85 ml/min.

herbicides flamprop-isopropyl and flamprop-methyl. The metabolism of [¹⁴C]CFA has been studied in the rat and dog as part of an industrial hygiene research programme. Excretion of metabolites in the two species was rapid and mainly via the urine. In a dog, 83% was eliminated in 48 h as the phenol sulphate, 2-amino-4-chloro-5-fluoro-phenyl sulphate, whereas the 0-24 h urine of rats after oral dosing contained the phenol sulphate (52%), N-(5-chloro-4-fluoro-2-hydroxy-phenyl)acetamide (13%) and an unidentified metabolite (16%). Only traces of radioactivity were detected in dog blood 24 h after dosing. In accord with a report [20] that dogs, unlike rats, do not N-acety-late aromatic amines, N-(5-chloro-4-fluoro-2-hydroxyphenyl)acetamide appeared to be absent in the dog.

The o- or p-position of anilines is the site of metabolic form-ation of conjugated aminophenols. The p-position in 3-chloro-4-fluoroaniline is blocked by fluorine; hence o-attack would be expec-ted. The 2-position is less accessible to attack than the 6-position because of the chlorine atom. The O-sulphate (which can be readily isolated as the cetylpyridinium salt) and the acetamide (Fig. 7) are both produced by substitution of the CFA in the 6-position. The O-sulphate may arise (i) by direct aryl hydroxylation followed by O-sulphation, (ii) via N-hydroxylation and rearrangement before or after sulphation, or (iii) via OH- attack on an N-hydroperoxide [21] followed by sulphation.

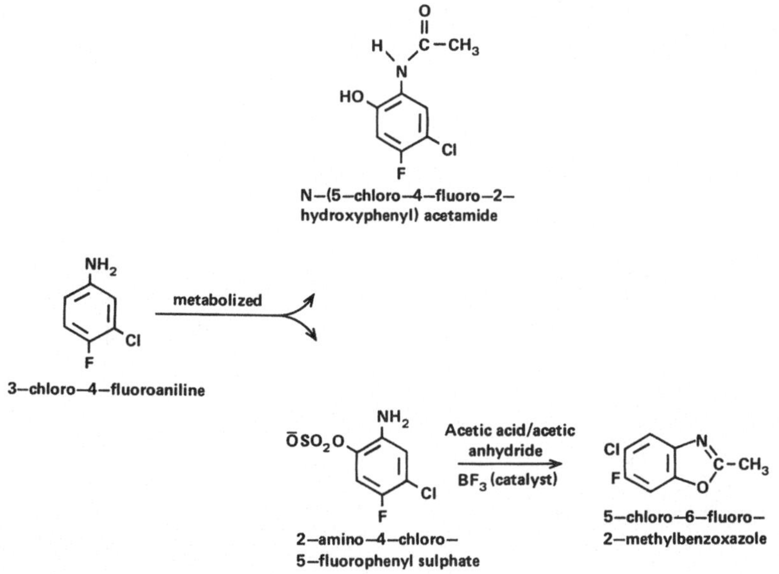

Fig. 7. Scheme for monitoring exposure to 3-chloro-4-fluoroaniline (CFA) using the metabolite 2-amino-4-chloro-5-fluorophenyl sulphate.

Information on CFA metabolism in man is lacking; but both the oxidative pathways ($C-$ and N-oxygenation) operate in man [22]. Also there is experimental evidence that human liver, unlike rat liver, N-acetylates aromatic amines only slowly [20]. The yield of the acetamide metabolite in man is thus likely to be less than for the rat. The phenol sulphate was therefore chosen for use in monitoring exposure of workers in the chlorofluoroaniline plant.

Previously the metabolite has been assayed after conversion into N-(5-chloro-4-fluoro-2-methoxyphenyl)formamide. The transformation was effected in two stages, firstly desulphation and N-formylation and secondly methylation of the resulting phenol. Recently a single-step conversion into 5-chloro-6-fluoro-2-methylbenzoxazole has been developed (cf. Fig. 7).–
(1) Urine dried down and derivatized with acetic acid/acetic anhydride/boron trifluoride (catalyst): reflux 3 h.
(2) Neutralize and extract twice with diethyl ether/hexane.
(3) Treat the organic phase with $KMnO_4/FeSO_4$ (oxidative clean-up).
(4) Analyze by GC-NPD (GC-AFID); confirm result by GC-MS.

ORGANOCHLORINE INSECTICIDES

Dieldrin

Dieldrin and endrin are two of a small group of chlorinated insecticides manufactured from hexachlorocyclopentadiene; others include chlordane, heptachlor, endosulfan, mirex and aldrin. Dieldrin and endrin are isomeric tetracyclic epoxyalkenes of formula $C_{12}H_8OCl_6$ (Fig. 8). Dieldrin is relatively stable in biological systems: hence

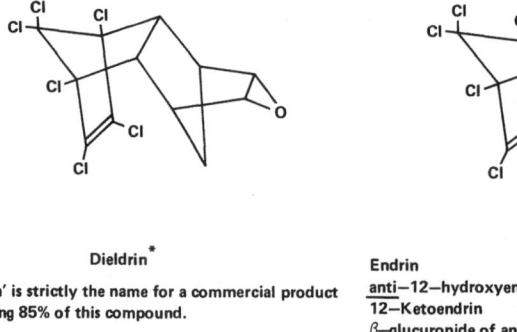

Dieldrin*

* 'Dieldrin' is strictly the name for a commercial product containing 85% of this compound.

	A	B
Endrin	H	H
anti—12—hydroxyendrin	H	OH
12—Ketoendrin		CO
β—glucuronide of anti—12—hydroxyendrin	H	D—glucuronic acid
anti—12—acetoxyendrin	H	Acetoxy

Fig. 8. Structures of dieldrin, endrin and endrin metabolites.

its content in body tissues such as fat and blood can be used as an indication of the body-burden on human beings, the previous intake level and the consequent hazard to health. The interrelationships as established by an ambitious kinetic study [23] allow meaningful monitoring of manufacturing, formulating and spraying operatives. The metabolism in man is slower than that observed [24, 25] in the common experimental mammals, but the major biotransformation pathway and elimination route is common to man and animals. Metabolite elimination is mainly in the faeces. Only 3% of the radioactivity from [^{14}C]dieldrin was eliminated in human urine after 5 days [1]. The level of dieldrin in blood may be determined as follows [26].-
(1) Prepare an acetone extract of blood, and clean up with a silica gel column.
(2) Add sodium sulphate solution, and extract with hexane.
(3) Chromatograph the organic phase on alumina with diethyl ether.
(4) Analyze the eluate by GC-ECD.

Endrin

Endrin (Fig. 8) is a dieldrin isomer and in mammals is metabolized much faster. This isomer effect has been explained in terms of the steric influence of the epoxide oxygen atom on C-12 hydroxylation. Hydroxylation at *syn*-C-12 is inhibited, so the metabolism of dieldrin is slow, and conversely for the *anti*-C-12 of endrin. In the rat the main metabolite is *anti*-12-hydroxyendrin (Fig. 8), which is eliminated in bile as its glucuronide but is excreted deconjugated mainly in the faeces [27]; 12-ketoendrin is a minor metabolite. In cows *anti*-12-hydroxyendrin is likewise the major metabolite but is found unconjugated in the urine [18].

In endrin-manufacturing operatives, endrin and metabolites were not detected in blood. Both endrin and *anti*-12-hydroxyendrin were found in faeces, but a method of analysis based on faeces was deemed undesirable, partly because of the difficulty of sample collection. The β-glucuronide of *anti*-12-hydroxyendrin, found as a urinary metabolite, seemed the best prospect for monitoring human exposure to endrin [29]. The available evidence suggests that man is similar to experimental animals in endrin metabolism.

The basis of an analytical method was mild periodate oxidation of the conjugate to give the formate ester of *anti*-12-hydroxyendrin (Scheme 1). This ester is labile and was cleaved in a carbonate buffer under mild conditions. The liberated metabolite was extracted and determined by GC-ECD without further purification. Oxidation to 12-ketoendrin or acetylation to *anti*-12-acetoxyendrin was used to confirm the results obtained with the periodate procedure.

Periodate results correlated well with β-glucuronidase cleavage results in urines from endrin plant workers. GC traces obtained for urine samples were of good quality, without appreciable interference

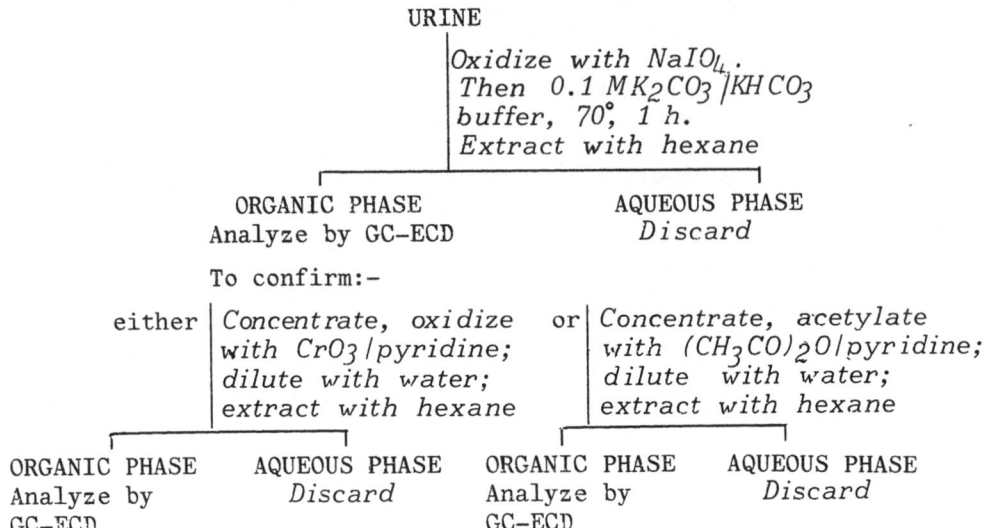

Scheme 1. Determination of the β-glucuronide of *anti*-12-hydroxy-endrin, a metabolite of endrin.

in the *anti*-12-hydroxyendrin peak position. There was, moreover, reasonable quantitative correlation between the original concentrations of *anti*-12-hydroxyendrin measured in the urine samples and the concentrations of 12-ketoendrin and *anti*-12-acetoxyendrin after oxidation or acetylation. This indicates that the peak originally ascribed to *anti*-12-hydroxyendrin was in fact correctly attributed.

References

1. Feldman, R.J. & Maibach, H.I. (1974) *Toxicol. Appl. Pharmacol.* *28*, 126-132.
2. Bull, D.L. & Lundquist, D.A. (1966) *J. Agric. Food Chem. 14*, 105-109.
3. Beynon, K.I., Hutson, D.H. & Wright, A.N. (1973) *Residue Rev. 47*, 55-142.
4. Shafik, T., Bradway, D.E., Enos, H.F. & Yobbs, A.R. (1973) *J. Agric. Food Chem. 21*, 625-629.
5. Shafik, M.T. & Enos, H.F. (1969) *J. Agric. Food Chem. 17*, 1186-1189.
6. Blair, D. & Roderick, H.R. (1976) *J. Agric. Food Chem. 24*, 1221-1222.
7. Ottenvanger, C.F. (1976) *An Epidemiological and Toxicological Study of Occupational Exposure to an Organophosphorus Pesticide*, Phoenix & Den Oudsten, Rotterdam, 188 pp.

8. Hutson, D.H., Akintonwa, D.A.A. & Hathway, D.E. (1967) *Biochem. J. 102*, 133-142.

9. Hutson, D.H., Holmes, D. & Crawford, M.J. (1976) *Chemosphere 5*, 79-84.

10. Hunter, C.G., Robinson, J., Bedford, C.T. & Lawson, J.M. (1972) *J. Occup. Med. 14*, 119-122.

11. Casida, J.E., Gaughan, L.C. & Ruzo, L.O. (1976) in *Advances in Pesticide Science (4th Internat. Congr. Pestic. Chem.*, Zurich, 1978), Vol. 2, Pergamon, Oxford, pp. 182-189.

12. Hutson, D.H. (1979) in *Progress in Drug Metabolism* (Bridges, J.W. & Chasseaud, L.G., eds.), Vol. 3, Wiley, Chichester, pp. 215-252.

13. Casida, J.E. & Ruzo, L.O. (1980) *Pestic. Sci. 11*, 257-269.

14. Gaughan, L.C., Unai, T. & Casida, J.E. (1977) *J. Agric. Food Chem. 25*, 9-17.

15. Hutson, D.H. & Casida, J.E. (1978) *Xenobiotica 8*, 565-571.

16. Gaughan, L.C., Ackerman, M.E., Unai, T. & Casida, J.E. (1978) *Agric. Food Chem. 26*, 613-618.

17. Crayford, J.V. & Hutson, D.H. (1980) *Pestic. Sci. 11*, 573-590.

18. Hirom, P.C., Millburn, P., Smith, R.L. & Williams, R.T. (1972) *Biochem. J. 129*, 1071-1077.

19. Baldwin, M.K. & Hutson, D.H. (1980) *Xenobiotica 10*, 135-144.

20. Glinsukon, T., Benjamin, T., Grantham, P.H., Weisburger, E.K. & Roller, P.R. (1975) *Xenobiotica 5*, 475-483.

21. Beckett, A.H. & Belanger, P.M. (1976) *Biochem. Pharmac. 25*, 211-214.

22. Parke, D.V. (1978) in *Drug Metabolism in Man* (Gorrod, J.W. & Beckett, A.H., eds.), Taylor & Francis, London, p. 61-69.

23. Hunter, C.G. & Robinson, J. (1967) *Arch. Environ. Health 15*, 614-626.

24. Hutson, D.H. (1976) *Food Cosmet. Toxicol. 14*, 577-591.

25. Müller, W., Nohynek, G., Woods, G., Korte, F. & Coulston, F. (1975) *Chemosphere 4*, 89-92.

26. Richardson, A., Robinson, J., Bush, B. & Davies, J.M. (1967) *Arch. Environ. Health 14*, 703-708.

27. Hutson, D.H., Baldwin, M.K. & Hoadley, E.C. (1975) *Xenobiotica 5*, 697-714.

28. Baldwin, M.K., Crayford, J.V., Hutson, D.H. & Street, D.L. (1976) *Pestic. Sci. 7*, 575-594.

29. Baldwin, M.K. & Hutson, D.H. (1980) *Analyst 105*, 60-65.

#NC(B)

NOTES and COMMENTS relating to

Investigation of metabolites, especially Phase I

Comments related to particular contributions:

#NC(B)-1

A Note on

METABOLIC FATE OF THE IMMUNOADJUVANT PEPTIDOGLYCAN MONOMER

Jelka Tomašić

"Ruđer Bošković" Institute
41000 Zagreb, Yugoslavia

Peptidoglycans are typical constituents of bacterial cell walls, with a basic structure in common [1]. They are composed of glycan chains, which are built of alternating *N*-acetylglucosamine and *N*-acetylmuramic acid residues, and peptide units which are made up of alternating L- and D-amino acids.

Both high and low mol. wt. peptidoglycan fragments exhibit versatile biological activities including immunostimulation [2, 3] and enhancement of tumour defence in experimental animals [4]. Peptidoglycans can be isolated from bacteria using physicochemical methods and enzymatic hydrolysis in combination. Low mol. wt. peptidoglycan fragments have been synthesized [5]. In 1975 we reported on a new method for obtaining non-crosslinked peptidoglycan fragments based on penicillin-induced inhibition of cell wall synthesis [6]. A biotin-requiring mutant of *Brevibacterium divaricatum*, when treated with penicillin, released peptidoglycan fragments into the culture medium. Upon treatment with lysozyme, monomer units were formed, isolated and chemically characterized as GlcNAc-MurNAc-L-Ala-D-iso-Gln-meso-di-aminopimelic acid-D-Ala-D-Ala [7]. By the use of ^{14}C-labelled precursor the peptodiglycan monomer (PGM) could be obtained labelled in both the disaccharide and the pentapeptide parts of the molecule.

Studies on biological activity have shown that PGM stimulates the humoral immune response to particulate antigens in mice [8, 9], inhibits the action of mitogens on mouse lymphocytes [10], and reduces the number of lung metastases in melanoma B-16 bearing mice (unpublished results).

Preliminary metabolic studies were carried out following i.v. administration of ^{14}C-PGM to mice. In the first 3 h, 60-80% of the

administered radioactivity was excreted in the urine, partly as unchanged PGM and partly as the corresponding pentapeptide L-Ala-D-iso-
Gln-meso-diaminopimelic acid-D-Ala-D-Ala [11].   Radioactive products
were separated on Sephadex G-25 (Fig. 1), and further purified by use
of Bio-Gels P-4 and P-2 and CM-Sephadex C-50.   The products were
characterized by hydrolysis and subsequent amino acid and hexosamine
analyses, dansylation, and comparison with authentic standards.

In vitro studies revealed that hydrolysis of PGM to the disaccharide and pentapeptide takes place in blood [12].   PGM remained
unchanged upon incubation with liver, spleen and kidney slices, but
was hydrolyzed in plasma and serum. The products in serum were isolated with Sephadex G-25 and further purified with Bio-Gels P-4 and P-2.
The disaccharide was characterized (a) by hydrolysis and hexosamine
analysis, (b) by reduction with NaBH$_4$ and subsequent analysis of glu-
cosamine and muraminitol, and (c) by confirming the presence of acetyl
groups in the molecule after hydrolysis as already described [7]. The
pentapeptide was characterized as described for the product from
urine, and the two products were compared chromatographically.

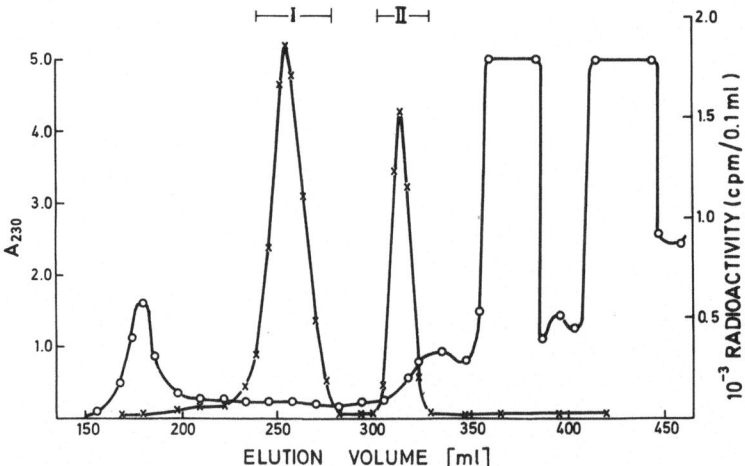

Fig. 1. Chromatography of urines excreted in the   first 3 h after
administration of ^{14}C-peptidoglycan monomer (PGM) to mice, each given
200 µg (21,780 cpm).   Concentrated urine samples (altogether 300,000
cpm) collected from 4 groups each of 5 mice were fractionated on a
95×2.5 cm Sephadex G-25 column with water as eluent.   The absorbance and radioactivity were determined on portions as indicated, and
radioactive fractions poooled as marked: I, unchanged PGM; II, newly
formed pentapeptide.   A$_{230}$ denoted o——o, and radioactivity ×———×.
From ref. [11].

Since PGM is hydrolyzed in blood, plasma and serum, evidently the enzyme N–acetylmuramyl–L–alanine amidase is present. It cleaves the lactylamide bond between N–acetylmuramic acid and L–alanine. An assay was developed using ^{14}C–PGM labelled both in the disaccharide and in the pentapeptide parts [13]. Following incubation the mixture was passed through Dowex–50W X4 in the H$^+$ form, and the disaccharide liberated during enzymatic hydrolysis was eluted with water. The measure of PGM hydrolysis was the recovered radioactivity. The amidase was partially purified by gel chromatography on Bio–Gel A–1.5 m, DEAE–Sephadex A–50 and Sephadex G–100 [13]. Purified amidase was used for preparing the disaccharide and pentapeptide in larger amount and for their final identification by NMR spectroscopy.

To conclude, peptidoglycan monomer is hydrolyzed in blood by the enzyme N–acetylmuramyl–L–alanine amidase to the corresponding disaccharide and pentapeptide subunits and excreted in urine very rapidly, partly as unchanged PGM and partly as the pentapeptide.

Acknowledgement

The author thanks Dr. D. Kelgević for a critical reading of the manuscript.

References

1. Schleifer, K.H. (1975) *Z. Immun.-Forsch. 149*, 104–117.
2. Heymer, B. (1975) *Z. Immun.-Forsch. 149*, 245–257.
3. Chedid, L., Audibert, F. & Johnson, A.G. (1978) *Prog. Allergy 25*, 63–105.
4. Lederer, E. (1980) *J. Med. Chem. 23*, 819–825.
5. Lefrancier, P. & Lederer, E. (1981) in *Progress in the Chemistry of Organic Natural Products* (Herz., W., Grisebach, H. & Kirby, G.W., eds.), Vol. 40, Springer–Verlag, Vienna, pp. 1–47.
6. Keglević, D., Ladešić, J., Hadžijka, O., Tomašić, J., Valinger, Z., Pokorny, M. & Naumski, R. (1975) *Eur. J. Biochem. 42*, 389–400.
7. Keglević, D., Ladešić, B., Tomašić, J., Valinger, Z. & Naumski, R. (1978) *Biochim. Biophys. Acta 585*, 273–281.
8. Hršak, I., Tomašić, J., Pavelić, K. & Valinger, Z. (1978) *Z. Immun.-Forsch. 155*, 312–318.
9. Hršak, I., Novak, D. & Tomašić, J. (1980) *Period. Biol. 82*, 147–151.
10. Hršak, I., Tomašić, J., Pavelić, K. & Benkpvić, B. (1979) *Period. Biol. 81*, 155–157.
11. Tomašic, J., Ladešić, B., Valinger, Z. & Hršak, I. (1980) *Biochim. Biophys. Acta 629*, 77–82.
12. Ladešić, B., Tomašić, J., Kveder, S. & Hršak, I. (1981) *Biochim. Biophys. Acta*, in press.
13. Valinger, Z., Ladešić, B. & Tomašić, J. (1982) *Biochim. Biophys. Acta*, in press.

#NC(B)-2

A Note on

HPLC AS A MEANS OF FOLLOWING AN *IN VITRO*
SYNTHESIS OF 7-HYDROXYMETHOTREXATE

Yahya Y.Z. Farid, Ian D. Watson,
Nigel G.L. Harding and Michael J. Stewart

Department of Biochemistry
Royal Infirmary, Glasgow G4 0SF, U.K.

Requirement	*Assay for methotrexate (MTX; a folic acid analogue) and its 7-hydroxy metabolite (7-OH-MTX) in liver-homogenate supernatants incubated with MTX.*
End-step	*RP ion-pair HPLC with detection at 307 nm.*
Sample preparation	*From incubation mixture take 1 ml, heat to 100°, and centrifuge at 2,500 rpm for 5 min. From supernatant take 20 µl for HPLC.*
Comments	*The assay allows 7-OH-MTX production to be monitored, and the kinetics of its formation from MTX to be measured. Substrate inhibition was noted.*

Excessive build-up of MTX in the patient leads to a syndrome of folate stress which, with the serum MTX level as a pointer, may be remedied by folinic acid therapy. The major metabolite of MTX is 7-OH-MTX which is much less water-soluble and may cause renal damage due to crystallization in the renal tubules [1]. Monitoring MTX and 7-OH-MTX is thus clinically relevant. We have recently developed an assay for MTX and its metabolites in serum (unpublished work) and detected what we believe to be 7-OH-MTZ behind the MTX peak. This position was atypical for an RP system; 3 mg of 7-OH were available to us and confirmed that the k' for 7-OH-MTX was greater than MTX, but was insufficient to pursue the question. An *in vitro* synthesis was therefore attempted, with homogenates (stored on ice till needed) of freshly killed rabbit liver prepared with an ATO-MIX homogenizer in 10 mM pH 7.6 Tris-HCl containing 0.25 M sucrose and 10 mM MgCl$_2$: from an 18,000 rpm supernatant (15 min, 4°) 20 ml was taken for incubation with 2-5 mg MTX. For monitoring and final isolation, see Fig. 1 legend.

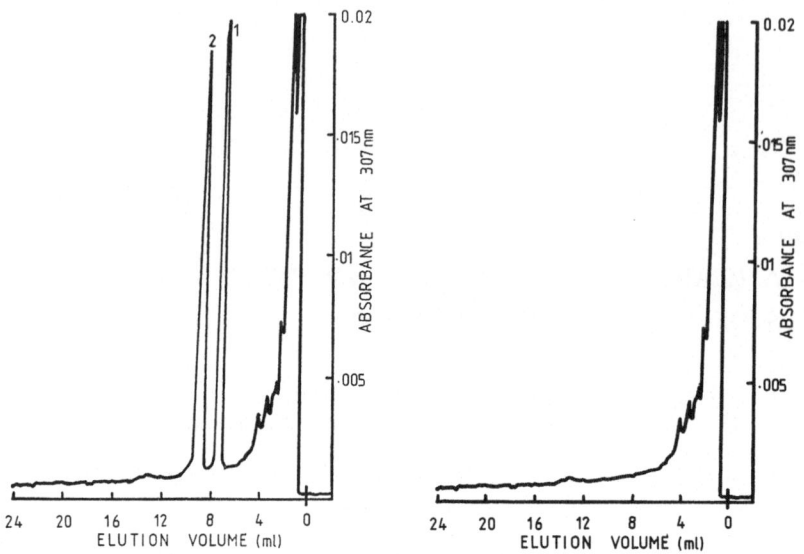

Fig. 1. HPLC on supernatants from liver homogenates. *Left:* 2.5 mg MTX/20 ml;　30 min at 37°. 1 = MTX; 2 = 7-OH-MTX. *Right:* blank run. Column: 100 x i.d. 5 mm, 5 µm Hypersil ODS.　Flow-rate 1.0 ml/min. Eluant: 50 mM phosphoric acid:methanol (72:28 by vol.),　containing 0.1% hexanesulphonic acid.　Detection: 307 nm.
　　Note on final isolation conditions: 16 x 5.2 cm column of DEAE-cellulose; elution in 500 x 10 ml fractions by 0.6 M ammonium bicarbonate, finally removed by freeze-drying after peak located.

Fig. 1 shows that the R_T for MTX was 7 min and for 7-OH-MTX was 9 min, and that unlike previous workers [2] we found no interferences.

The conversion of MTX into 7-OH-MTX was concentration-dependent, being maximal (usually ~100%) after 1 h with 0.05 mg/ml but not till 5 h with 0.5 mg/ml.

Acknowledgement

Dr. David G. Johns (National Cancer Institute, Bethesda, MD) is thanked for the gift of 7-OH-MTX.

References

1. Jacobs, S.A., Staller, R.G., Chabner, B.A. & Johns, D.G. (1976) *J. Clin. Invest. 57,* 534-538.
2. Watson, E., Cohen, J.L. & Chan, K.K. (1978) *Cancer Treatment Repts. 62,* 381-387.
Added by Editor: a comparable RP-HPLC assay for the same analytes, on deproteinized serum –

3. Lawson, G.J. & Dixon, P.F. (1981) *J. Chromatog. 223,* 225-231.

Comments on material in #B

Background to the study of metabolic pathways: some points made by J.W. Bridges (University of Surrey, Guildford)

The concept that observed biological effects should be directly related to the levels and/or persistence of a drug in target tissues has led some authorities to recommend that species selection for toxicology studies should aim at a urinary metabolite profile like that of man. Valid experimentation entails problems.-
1. Animal models have to be the basis for restricted studies in man.
2. Drug and metabolite determinations on plasma and other biological fluids have to serve, by extrapolation, for assessing the extent and duration of interaction with the receptor, even if its nature has been elucidated.
3. A drug effect may be due largely to reactive intermediates which are difficult to assay, are prone to destruction when common isolation procedures are applied, and readily bind to diverse cellular constituents, few of which appear crucial to the observed biological effect(s).
4. The level, duration and route of dosage and the presence of other drugs may each profoundly affect which pathway predominates.
5. Knowledge of possible metabolic pathways is as yet incomplete.

B. Scales, *commenting on remarks by* H. de Bree.- There is indeed a very poor understanding of what levels of metabolites call for further investigation of their structure and pharmacokinetic profile. Some regulatory authorities demand a fairly thorough study of 'major' metabolites, and ensure that this is done even though the one major metabolite accounts for only, say, 0.2% of the dose (the rest being unchanged drug) and has no therapeutic or toxicological effect. A metabolite really merits study only if it has a therapeutic effect or a toxic effect, and if it occurs not only in the urine (or other excreta) but also in the blood or tissues where an effect can be elicited. *Remark by* H. Ritter, concerning drug elimination in rodents *vs.* man.- With several of our drugs, elimination via bile into faeces accounts for more than 50% of the oral dose.

Comments on #B-1, G.R. Bourne - INVESTIGATING METABOLITES (Phase I)

Replies to J.A.F. de Silva and J. Vessman.- The ^{14}C-labelling of the prostaglandin analogue was 98% of theoretical, so 'hot' as to

pose a stability problem.   The preparation was made for a single
high-dose experiment in the cow; further work was done with pre-puri-
fied crystalline material.   In the tissue-clearance and other studies
on cloprostenol, ^{14}C was a better choice than a stable isotope such
as ^{13}C: the radiolabelled material was readily detected in tissues,
after appropriate sample preparation, by liquid scintillation counting, and the high s.a. that had to be used enabled us to develop
the ion-cluster technique for metabolite identification.   Such use
of this technique is more commonly associated with compounds having
a stable-isotope label, which for the present range of work would
have had only limited applicability.

G.R. Bourne's *replies to questions by* H. de Bree, J. Chamberlain
and B.F.H. Drenth.- To dry down HPLC fractions for NMR, organic components are removed by rotary evaporation, and the remaining water by
lyophilization.   The amount of material we seek for NMR, to allow
good interpretation by a spectroscopist, is 10-20 mg, although we
had much more in the present study but, on the other hand, 10 µg
is theoretically feasible. The choice amongst possible types of evidence in metabolite identification (e.g. retention time, NMR, MS) is
governed by the need to rely on physico-chemical data for the compounds where our experienced medicinal chemists are unable to syn-
thesize them and we rely, for their 'synthesis', on our animals or
animal model systems.   The validity of inferring *in vivo* metabolism
from results with isolated perfusion systems is evident from the
example cited (in #B-1), where the drug was 100% metabolized: the
perfusion patterns were identical with those found *in vivo* (bile
and faeces).

Comments on #**B-2**, I.D. Watson et al.- AMITRIPTYLINE, BENZODIAZEPINES

Queries and comments by R.G. Muusze.- Was the level of
10-hydroxyamitriptyline in blood ascertained? (*Reply*: no!)   What
was the basis for the diagnosis 'endogenous depression'? (*Reply:* use
by the patient of only one antidepressant.) With an NP system using
acetonitrile/water/ammonia, 10-hydroxyamitriptyline and nortriptyline are readily separable. *Suggestion by* A.S. Papadopoulos.- TLC
in place of GC, to separate metabolites before MS examination, might
have circumvented the problem of phthalate derivatives in the column
eluate.   I have found TLC a simple means of separating unidentified
thioridazine metabolites.

In answer to J.A.F. de Silva.- Whereas EMIT gives fast, reliable *identification* of all classes of drug involved in an overdosage,
HPLC serves as follow-up for confirmation and for estimating, at
least semi-quantitatively, the particular drug/metabolite(s).   For
flurazepam (*remark by* J. Ramsey), the EMIT approach might have to
be used cautiously in patients who have steady-state levels of des-
alkylflurazepam. *Note on radioreceptor assays - SEE BELOW.*

Identification, and procuring authentic material: further points

Remarks by W. Ritter.- Our philosophy is that metabolites can be identified only by comparing the mass spectra of the isolated metabolite and the synthesized metabolite. Otherwise identification has to be termed "likely" or "with some evidence".

Specimens for psychotropic drug studies.- In relation to the plea for a central source which concludes art. #B-2, it is a boon to the analytical fraternity that specimens of numerous compounds (including tyramine derivatives and ß-endorphin) - with an impressive range of metabolites - can be sought for a stated need from the Neurosciences Research Branch (Dr. Stephen Kennedy) of the National Institute of Mental Health, Rockville, MD 20857, USA (list available). *Steroids* are held by Ref. Collection Curator, Westfield Coll., London NW3.- *Ed.*

Comment on #B-4, C.V. Eadsforth - PESTICIDE METABOLITES

Remark by U.A.Th. Brinkman.-Many of the compounds you studied seem to be well suited for analysis by HPLC instead of GC; would UV detection suffice ? *Reply.-* I am sure that similar detection limits could be achieved for the wild oat herbicide and 3-chloro-4-fluoroaniline metabolites with UV detection (or electrochemical detection). Most of the biological methods that have been developed in the last year in my laboratory have utilized HPLC (UV detection).

Comment on #NC(B)-1, J. Tomašić - PEPTIDOGLYCAN IMMUNOADJUVANT

Answer to J. Rosenthaler.- It is indeed only the pentapeptide portion that is immunologically active and not the saccharide portion. The reaction mechanism is unknown, but may well involve the macrophages as might be expected. *In response to* I.D. Watson: In exerting its antineoplastic action, most effectively against metastases (ineffective against the primary neoplasm), peptidoglycan may be relating to differences in cell-surface structure and properties; but the mechanism remains to be elucidated.

Some general points concerning metabolites

Example of artefact troubles (J.W. Gorrod).- A nicotine metabolite, cotinine, found in the urine of smokers has to be assayed, if an accurate result is to be obtained, immediately after urine collection because cotinine is also an artefact of pyridylbutyric acid (ring closure reaction). *Follow-up thoughts by* W. Ritter.-One wonders what happens in the bladder, and whether cotinine is a metabolite at all: is it a *metabonate* (artefact) ? If primary metabolites in urine are unstable and are transformed to metabonates *outside* the kidney but *inside* the body, i.e. in the bladder, what then is the definition of a metabolite ?

Biliary excretion: background refs. (Ed.) – An introductory account published in 1973 [1] remains useful for gaining perspective. A recent account [2] of hepatobiliary disposition goes deeply into the subject, with comprehensive listing of data for xenobiotics (not merely drugs) in different species.

1. Smith, R.L. (1973) *The Excretory Function of Bile*, Chapman & Hall, London, 394 pp.
2. Klaassen, C.D., Eaton, D.L. & Cagen, S.Z. (1981) in *Prog. Drug Metab. 6*, 1-75.

Comment on #NC(B)-2, Y.Y.Z. Farid et al. – INCUBATE MONITORING

Query by H. de Bree.– Were the experiments done with the racemate or one of the enantiomers ? – *Reply:* Only commercially availble methotrexate was used.

Radioreceptor assay for benzodiazepines – Editorial note [cf.#B-2]

With receptor material from brain, there is specificity for benzodiazepines; all are responsive [1], including active metabolites (Amersham International leaflet, 1982).

1. Braestrup, C. & Squires, R.F. (1978) *Eur. J. Pharmacol. 48*, 263-270.

N-dealkylation of tertiary amines: any enzyme available as a tool? *(Bibliographic search by E. Reid)*

It is difficult to derivatize a tertiary amino group, as may be desired for assay purposes where volatility and/or UV detectability is sought. An enzymic means of removing alkyl group(s), without side reactions, would be a boon in this connection (e.g. with formaldehyde estimation as the end-step), and also for identification purposes or for preparing authentic material. There is, however, little encouragement in the literature, which is centred on liver from different species and especially on microsomal fractions [e.g. 1].

To summarize, *N*-demethylation activity is, in the crude preparations that dominate the literature (containing NADPH-cyt.*c* reductase), found only accompanied by *N*-oxidase activity. Formaldehyde of alkylamino (sec.) origin turned out to be an acid degradation product of enzymically generated *N*-methyl-alkylhydroxylamine [1].

1. Ziegler, D.M. & Mitchell, C.H. (1972) *Arch. Biochem. Biophys. 150*, 116-125.

Section #C

INVESTIGATION OF CONJUGATES

#C-1

ISOLATION OF CONJUGATES AND IDENTIFICATION

OF THE ATTACHED GROUPS

Jelka Tomašić

"Ruder Bošković" Institute
41000 Zagreb
Yugoslavia

Metabolic Phase II reactions convert a wide variety of drugs, endogenous substances and environmental chemicals into conjugates. The application of various isolation and characterization procedures to the analysis of several types of conjugates is considered, including:
(a) conjugates with glucuronic acid, special attention being paid to some novel types, to possible formation of isomeric structures in ester-type glucuronides, and to the use of ß-glucuronidase in the assay of glucuronides;
(b) conjugates with amino acids, and particularly glutathione conjugates and mercapturic acids, mention being made of substrate specificity of glutathione S-transferase;
(c) sulphoconjugates as products of conjugation of steroids and phenols.

Since several types of conjugation reaction often compete for the same substrate, especially in steroid series, possibilities for group separation of various conjugates are touched on.

Isolation of conjugates is normally hampered by the presence of relatively large amounts of naturally occurring compounds such as proteins, lipids, carbohydrates, hormones, amino acids and salts, and on the other hand by relatively low concentrations of particular metabolites in unconcentrated physiological fluids and faeces. [This situation is explored in other articles - notably #A-2 - together with the nature of approaches such as HPLC.- *Ed.*]

The choice of method for isolation depends on the stability of the parent compound and the respective metabolites, on specificity and accuracy requirements, and on the instruments and chemicals

available. Since conjugation is usually the final step in metabolism
the conjugates are excreted either in urine or in bile and faeces,
maybe with minor amounts in blood.   Numerous metabolic studies are
also performed *in vitro*.   Consequently, methods for isolating and
characterizing conjugates should be applicable to each of the above-
mentioned biological materials.   As others emphasize (e.g. # A-2), the
use of radiolabelled compounds is very advantageous; otherwise a very
specific assay for the parent compound and related metabolites should
be provided in order to distinguish them from interfering endogenous
substances.

   Whilst metabolic transformation is best established by isolating
intact metabolites including conjugates, chemical or enzymic hydrolysis is often adopted as a speedier alternative; liberated constituents are assayed separately.   This provides general information on
the type of conjugation,   but in some cases might preclude ascertaining simultaneously the relative abundance of individual conjugates
and the attachment position of the conjugating group.

   It is worth starting the isolation procedure by extracting non-
polar metabolites and unchanged parent compound. Conjugates remaining in the aqueous phase may then. be concentrated by XAD resins, or
directly analyzed chromatographically.   Acidic conjugates may also be
extracted after acidifying the aqueous layer or saturating it with
inorganic salts, e.g. ammonium sulphate.   Chromatographic purification
may be complete, or partial, with final purification by more sophisticated instrumental methods.

   Since hundreds of reports on conjugate isolation and characterization have appeared in the last few years, the following survey is
confined to a few examples for each type of conjugate.

CONJUGATES WITH GLUCURONIC ACID

   Glucuronide formation, a major route in the metabolism of drugs
and endogenous compounds, leads to several classes of conjugates [1,
2] differing considerably in their properties and stability in alkaline and acidic media – as should be considered when choosing a method
for isolation and characterization [3].   The group-specific enzyme
β-glucuronidase [4] has been an indispensable tool; it hydrolyzes only
conjugates with a β-D-configuration and pyranoside structure.   Some
N-glucuronides and *C*-glucuronides are not attacked. Hydrolysis with
β-glucuronidase has been considered as evidence for the glucuronide
structure of metabolites, and in many reports has remained the only
evidence. Hydrolysis is strongly inhibited by D-glucaro-1,4-lactone.
Parallel experiments with the inhibitor minimize the possibility of
obtaining false results due to impurities in enzyme preparations or
chemically induced changes in the course of isolating and handling
the sample.

Hydrolysis is usually monitored by a change in chromatographic behaviour after treatment with β-glucuronidase or by extraction and assay of liberated aglycone. TLC or paper-chromatographic redistribution of the label is a reliable parameter for quantitative studies. Several metabolic studies based primarily on hydrolysis with β-glucuronidase and simple chromatographic assays have been described recently: metabolism of lormetazepam [5, 6], flavanones [7], 3-cyano-4,5-dimethyl-2-pyrrolyloxamic acid [8], 2-aminophenol [9], metronidazole [10], eleven diverse aglycones including steroids, phenols and aromatic acids studied in respect of metabolism [11], oestradiol [12], etc.

Glucuronide isolation and separation can sometimes be carried out by preparative TLC and paper chromatography, as reported by Parker et al. [13] for conjugates of phenolphthalein, LSD, morphine and diphenylacetic acid from bile and faeces. The glucuronide of 3'-O-methyl-(+)-catechin from bile was isolated by preparative paper chromatography [14], and the glucuronide of 3-OH-benzo(a)pyrene was isolated from incubation mixtures by preparative TLC and assayed fluorimetrically [15].

More elaborate methods including derivatization of isolated aglycones or intact glucuronides have also been reported. In the study of LSD metabolites [16], glucuronides of 13- and 14-hydroxy-LSD were isolated from bile and from liver perfusates using solvent extraction, concentration on XAD-2 and preparative TLC; following hydrolysis with β-glucuronidase, aglycones were characterized by MS and derivatization. A similar approach was used for the isolation and assay of aniline mustard metabolites from bile [17].

HPLC has been extensively studied and applied for separating intact glucuronides. Thus, metabolites of 13-cis-retinoic acid from bile [18] were first separated by HPLC and the glucuronides detected subsequently by hydrolysis with β-glucuronidase. Liberated aglycones were characterized by comparison of chromatographic parameters with those of synthetic standards. Metabolites of zinc pyridinethione were separated by direct HPLC of urine samples [19]. Identification of two structurally different S-glucuronides was carried out by comparison of ^{13}C-NMR and IR spectral data with those of synthetic standards, by derivatization of conjugates and subsequent MS, and by reverse isotopic dilution analysis.

In some cases partial purification and isolation of conjugates from very complex mixtures might be required prior to final purification by HPLC, as shown for metabolites of the analgesic agent zomepirac sodium [20], metabolites of the anti-inflammatory fendosal [21] and metabolites of vitamin D_2 [22]. Zomepirac glucuronide was extracted from urine and chromatographed on Sephadex LH-20. The conjugate of hydroxylated fendosal was extracted from faeces with methanol and partially purified by chromatography on silica gel.

Charged metabolites of the 25-hydroxy derivative of Vit. D_2 were isolated by chromatography on DEAE-Sephadex A-25 and further purified by chromatography on Sephadex LH-20. The structures of the above-mentioned conjugates and the position of attachment of the glucuronyl moiety were confirmed by a combination of IR, NMR, GC, MS and hydrolysis with β-glucuronidase.

Steroids are mostly excreted as glucuronides, but sulphoconjugation and conjugation with glucose and N-acetylglucosamine are consistent although sometimes quantitatively minor pathways in the metabolism of steroids. The procedures applied in the study of steroid metabolism always include the group separation of various types of conjugates, and are discussed in a later section.

α-Anomers and conjugates in furanose forms have not been detected in biological materials until very recently. It has also been assumed that the aglycone is always attached to the C-1 position of glucuronic acid. In several recent reports the occurrence of various structural forms of glucuronic acid conjugates has been claimed, especially in the series of 1-O-acyl glucuronic acids. Such structures can arise from intramolecular acyl migration and structural rearrangements in the molecules of partially acylated conjugates. Such processes are well documented and have been studied with synthetic model compounds [23, 24]; they can be both acid- and base-catalyzed and are liable to occur even under mildly alkaline conditions [25] or in the course of methylation with diazomethane [26]. Illing & Wilson [27; cf.#C-3] demonstrated the occurrence of several glucuronic acid conjugates of isoxepac in urine using HPLC; the number of peaks was dependent in part on the age of the sample. New structural isomers were found as metabolites of probenecid [28]. Following separation by HPLC, conjugates were tentatively characterized as anomers of probenecid glucuronide, in which the aglycone was not linked to glucuronic acid via the C-1 position. In this work ^{13}C-NMR was used for characterization.

Four different glucuronic acid conjugates were found as urinary metabolites of clofibric acid [29]. Conjugates were separated by GC after appropriate derivatization, and characterized on the basis of MS data as α- and β-anomers in both furanose and pyranose forms. Altogether, recent findings present a strong indication that isomers and anomers of glucuronic acid conjugates could be present in biological material, and hence should be looked for in future metabolic studies.

MERCAPTURIC ACIDS AND CONJUGATES WITH GLUTATHIONE

The principal pathway for converting many electrophilic agents, including arenes, olefins and halogenated hydrocarbons, into excretable metabolites is conjugation with glutathione. In vivo, it can be followed by formation and excretion of mercapturic acids, i.e.

conjugates with N-acetylcysteine. Both types of thioether conjugates
have been extensively studied *in vivo* and *in vitro*. Xenobiotic arenes
and olefins are first converted into epoxide intermediates in Phase I
reactions, followed by nucleophilic attack of the sulphydryl group of
glutathione at the electrophilic atom on the hydrophobic substrate.
Since the epoxide ring can be opened in two different ways, the
formation of two positional isomers can be expected. Moreover, the
introduction of a new asymmetric centre can give rise to two diastereoisomers, thus making possible a total of four different forms
of each metabolite.   This should be taken into account when choosing
appropriate methods for isolating glutathione conjugates and mercapturic acids.

The development of new high-resolution chromatographic techniques, especially HPLC, has made the separation of isomers possible.
Simpler chromatographic methods, e.g. preparative TLC and conventional
column chromatography, have also been used for isolating such conjugates. The evidence for amino acid conjugate formation has usually
been based on the assay of constituent amino acids. Most frequently,
identification of conjugates has been carried out by comparison with
synthetic standards.

Nemoto and Gelboin [30] used TLC for isolating and characterizing
the conjugate of benzo(a)pyrene-4,5-oxide with glutathione, and
Chassaud et al. [31] for the conjugate of benz(a)anthracene-5,6-oxide.
Urinary mercapturic acids formed from acrylate, methacrylate and crotonate were extracted, methylated and isolated on silica-gel columns
[32]; metabolites were characterized by comparing chromatographic
parameters with those of synthetic standards. Similar procedures were
applied for isolating mercapturic acids formed from acrylonitrile,
crotononitrile and cinnamonitrile [33]; conjugates were identified
by MS and NMR spectroscopy.

Urinary mercapturic acids formed from cyclohexane oxide [34]
were likewise isolated by column chromatography on silica gel.
Quantitative determination of diastereoisomers was performed by GC
and identification on the basis of NMR spectra. Two isomeric mercapturic acids were isolated from urine following administration of
styrene [35]. After isolation of methylated conjugates on silica-gel
columns, final purification was achieved by preparative TLC. Metabolites were characterized by IR, NMR and MS. Diastereoisomers were not
separated but were detected in the NMR spectra of each isomer.

Watabe et al. [36] studied glutathione conjugates of styrene
oxide, phenyloxirane, *in vitro*. Two isomeric conjugates were isolated
using XAD-2, separated by HPLC and examined for identity by comparison
with synthetic compounds. Structures were assigned on the basis of
NMR spectra. In the study of benzo(a)pyrene-4,5-oxide metabolism
[37], biliary metabolites were separated by HPLC without prior clean-
up. Thioether conjugates were identified by hydrolysis and subsequent

assay of constituent amino acids. The presence and partial separation
of isomeric glutathione conjugates and also of a cysteine conjugate
was reported. The stereoselectivity of glutathione S-transferase
toward aralkyl halide enantiomers was studied in vitro [38], and
preference of the enzyme for (S)-phenetyl chloride established. The
conjugates were isolated by the use of XAD-2, separated by HPLC, and
identified by comparison with synthetic standards.

Recently van Doorn et al. [39] reported finding o-methylbenzyl-
mercapturic acid as a urinary metabolite of o-xylene and benzylmer-
capturic acid as a metabolite of toluene. Conjugates were isolated
from acidified urine by extraction, purified by crystallization and
identified by MS and NMR.

Examples of the isolation of glutathione conjugates in parallel
with sulphates or glucuronides are given in a later section.

SULPHOCONJUGATES

Sulphoconjugation often accompanies glucuronidation, but in
particular physiological states or under specially adjusted conditions
in vitro it can be markedly increased. A common property of sulphoconjugates is their sensitivity to acid hydrolysis. A mild hydrolytic procedure has been reported [40] for sulphates in an appropriate
non-polar medium or in the aqueous phase accompanied by solvent
extraction, and has been widely used in analyses of this type of conjugate. In the solvent-extraction of metabolites from acidified urine
or physiological fluids, special care should be taken to avoid
unwanted solvolysis of sulphates. Hydrolysis with the group-specific
enzyme sulphatase is very often used as evidence for sulphoconjugate
structure. However, some sulphoconjugates are not cleaved with
particular enzyme preparations or are not susceptible to enzymic
hydrolysis at all.

Various chromatographic methods have been applied in sulphoconjugate analyses. Goodall and James [41] tentatively characterized
an androstenedione conjugate from plasma as the 3-enol sulphate. The
conjugate was isolated by TLC and characterized by hydrolysis with
sulphatase, solvolysis and comparison with synthetic standards. 2-
Amino-4-chloro-5-fluorophenyl sulphate [42] was isolated from urine
using XAD-2, purified as the potassium or cetylpyridinium salt by
recrystallization, and identified by MS and NMR. Sulphated steroids
from faeces [43] were extracted and subsequently purified by chromatography on Sephadex LH-20. Following solvolysis and derivatization,
bile acids and cholesterol were identified by GC-MS. Bell et al. [44]
demonstrated the use of high-voltage paper electrophoresis for isolating and characterizing drug metabolites, including the sulphate
ester of salbutamol.

Most recent researches are based on HPLC, which can provide a

distinct separation of metabolites. Crayford and Hutson [45] characterized 3-(4-sulphonyloxyphenoxy)benzoic acid as a major metabolite
of 3-phenoxybenzoic acid in urine. Metabolites were separated by HPLC
and characterized by hydrolysis with sulphatase and TLC. Takeishi et
al. [46] found two new metabolites of isopropylantipyrine from urine,
and characterized them as two·isomeric enol sulphates. Following
concentration on XAD-2, metabolites were first purified by preparative TLC and subsequently separated by HPLC; they were identified by
solvolysis and chemical characterization of the products. The 3-
sulphate isomer was shown to be the major product of sulphation of
chenodeoxycholate conjugates *in vitro* [47]. The incubation medium
was first subjected to preparative TLC, and potential sulphated
products were purified by HPLC. The sulphate conjugate was also found
as a metabolite of 3-methoxy-4-hydroxyphenylethylene glycol in human
CSF [48]. Non-polar metabolites were extracted from CSF, and the
remaining aqueous layer submitted to hydrolysis with sulphatase.
Liberated parent compound was assayed by HPLC.

PARALLEL ISOLATION OF DIFFERENT TYPES OF CONJUGATES

In the course of metabolic transformations, several types of
conjugates could be formed from one parent compound. In view of the
growing evidence that the presence of certain types of conjugates
indicates a particular physiological state, especially in the steroid
area, methods providing complete metabolic profiles are required. An
essential step in such an analysis is the group separation of the
conjugates involved. The use of radiolabelled compounds considerably
facilitates such complex metabolic studies and is of particular help
in following the course of fractionation. In order to obtain information concerning the nature of conjugation, sequential hydrolysis
with β-glucuronidase and sulphatase is almost invariably applied.

Steroid metabolism usually leads to very complex mixtures of
non-conjugated metabolites, conjugates and di-conjugates. Hence the
methods for isolating, separating and characterizing steroid metabolites have received a great deal of attention and have been continu-
uously modified and improved. Until recently, the possibilities for
separation of steroid and bile acid metabolites comprised various
combinations of TLC, paper chromatography, gel and ion-exchange
chromatography, as shown in several examples now given.

Metabolites of oestradiol and ethynyl oestradiol [49] were
isolated from bile by preparative TLC, which afforded free steroids,
sulphates and glucuronides. Metabolites of synthetic oestrogens,
administered as diethylstilboestrol sulphates, were separated by TLC
into sulphates, glucuronides and di-conjugates [50]. Samarajeeva et
al. [51] elaborated the procedure for isolating oestriol-16α-glucuronide
and pregnanediol-3α-glucuronide from pregnancy urine. Group separation of sulphates and glucuronides was carried out by chromatography
on deactivated alumina. Final purification of the glucuronides was

achieved by partition chromatography on celite. Urinary metabolites of diethylstilboestrol [52] were also separated on an alumina column.

A detailed analysis of bile acid conjugates, particularly glucuronides, from urine was reported by Almé and Sjövall [53]. Following preliminary purification on XAD-2, the separation of conjugates was carried out on the lipophilic ion-exchanger DEAP-Sephadex LH-20. Since only partial separation was achieved, the mixture consisting of glucuronides, taurine conjugates and sulphates was methylated and re-chromatographed on the same ion-exchanger to yield methylated glucuronides in neutral fractions and sulphates in charged fractions. Individual glucuronides were finally separated by GC and characterized by MS.

The possibility of using a so-called liquid ion-exchanger for paper chromatography of steroidal glucosiduronic esters was reported by Mattox and Litwiller [54]. (This approach represents 'ion-pair extraction'.- *Editor*.) The metabolism of androsterone and the separation of biliary metabolites were studied by Matsui and Aoyagi [55]. Group separation of glucuronides and sulphates, after purification on XAD-2, was carried out by chromatography on Sephadex LH-20.

Rapid and efficient resolution of free and conjugated metabolites of oestradiol from urine and plasma was achieved by HPLC [56]. After removal of salts and proteins by precipitation with organic solvents, the remaining material was subjected directly to HPLC. By appropriate combination of several systems, resolution of non-conjugated and di-conjugated oestrogens was achieved. The amounts of several metabolites were also determined on the basis of recovered radioactivity.

Different conjugate types from foreign compounds

In principle, the methods used for analysis of steroid conjugates can be applied to the diverse types that arise from many foreign compounds. TLC in combination with enzymic hydrolysis was used for separation of glucuronides and sulphates as metabolites of phenol [57], carbaryl [58], acetaminophen [59], carfecillin [60] and several xenobiotics including phenol, naphthalene, benzoic acid and aniline [61]; also aspirin, *p*-aminobenzoic acid and sulphadiazine [62]. Sulphates, glucuronides and glutathione conjugates found as *in vitro* metabolites of benzo(a)pyrene [63] were separated by column chromatography on neutral alumina. Fractions containing the glutathione conjugates were further purified by HPLC and resolved into several components.

Sulphates and glucuronides of aminostilbene [64] were isolated using alumina followed by chromatography on Sephadex LH-20, as also used for separating various acidic conjugates of dimethylaminophenol [65, 66]. Paper electrophoresis combined with Sephadex LH-20 chromatography was applied in the isolation of acidic conjugated urinary metabolites of o-toluidine [67].

The metabolism of paracetamol was studied from various aspects using HPLC [68-70], which provided one-step resolution of glucuronide, sulphate and glutathione conjugates or mercapturic acids. The glucuronide and sulphate of naphthalene were also resolved by HPLC and identified on the basis of retention times [71].

CONCLUDING COMMENTS

In the last few years the approach to the isolation and analysis of acidic conjugates has changed considerably, the emphasis being put on instrumental techniques, particularly HPLC. The use of high-performance chromatographic techniques enabled in some cases the detection and isolation of new isomers, anomers or diastereoisomers from biological material. However, simple chromatographic methods have also been used for conjugate analysis and proved to be satisfactory, especially in preliminary metabolic studies. In appropriate combination with enzymic hydrolysis and comparison with synthetic standards, such simple methods can be satisfactory even for complete metabolic studies.

Acknowledgement

The author thanks Drs. D. Keglević and S. Tomić for a critical reading of the manuscript.

References

1. Williams, R.T. (1959) *Detoxication Mechanisms*, 2nd edn., Wiley, New York.
2. Fishman, W.H. (1961) *Chemistry of Drug Metabolism*, C.C. Thomas, Springfield, Illinois.
3. Tomašić, J. (1978) in *Drug Fate and Metabolism*, Vol. 2 (Garrett, E. & Hirtz, J.L., eds.), Dekker, New York, pp. 281-335.
4. Wakabayashi, M. (1970) in *Metabolic Conjugation and Metabolic Hydrolysis*, Vol. 2 (Fishman, W.H., ed.), Academic Press, New York, pp. 519-602.
5. Hümpel, M., Illi, V., Milius, W., Wendt, H. & Kurowski, M. (1979) *Eur. J. Drug Metab. Pharmacokinet. 4*, 237-243.
6. Mayo, B.C., Hawkins, D.R., Hümpel, M., Chasseaud, L.F. & Girkin, R. (1980) *Xenobiotica 10*, 413-420.
7. Hackett, A.M., Marsh, I., Barrow, A. & Griffiths, L.A. (1979) *Xenobiotica 9*, 491-502.
8. Honigberg, I.L. & Meltzer, N.M. (1981) *Drug Metab. Dispos. 9*, 67-68.
9. Burchell, B. & Bock, K.W. (1980) *Biochem. Pharmacol. 29*, 3204-3207.
10. LaRusso, N.F., Lindmark, D.G. & Müller, M. (1978) *Biochem. Pharmacol. 27*, 2247-2254.
11. Bansal, S.K. & Gessner, T. (1980) *Anal. Biochem. 109*, 321-329.

12. Zumoff, B., Freed, S.Z., Levin, J., Whitmore, W.F., Hellman, L.,
Fishman, J. & Fukushima, D.K. (1980) *Eur. J. Cancer 16*, 219–221.

13. Parker, R.J., Hirom, P.C. & Millburn, P. (1980) *Xenobiotica 10*,
689–703.

14. Shaw, I.C. & Griffiths, L.A. (1980) *Xenobiotica 10*, 905–911.

15. Bock, K.W. & Lilienbaum, W. (1979) *Biochem. Pharmacol.   28*,
695–700.

16. Siddik, Z.H., Barnes, R.D., Dring, L.G., Smith, R.L. & Williams,
R.T. (1979) *Biochem. Pharmacol. 28*, 3081–3091.

17. Chipman, J.K., Hirom, P.C. & Millburn, P. (1980) *Biochem. Pharmacol. 29*, 1299–1301.

18. Frolik, C.A., Swanson, B.N., Dart, L.L. & Sporn, M.B. (1981) *Arch.
Biochem. Biophys. 208*, 344–352.

19. Jeffcoat, A.R., Gibson, W.B., Rodriguez, R. A., Turan, T.S.,
Hughes, P.F. & Twine, M.E. (1980) *Toxicol. Appl. Pharmacol. 56*,
141–154.

20. Wu, W.N., Weaner, L.E., Kalbron, J., O'Neill, P.J. & Grindel,
J.M. (1980) *Drug Metab. Dispos. 8*, 349–352.

21. Warrander, A., Metcalf, R. & Fromson, J.M. (1981) *Drug Metab.
Dispos. 9*, 161–167.

22. Le Van, L.W., Schnoes, H.K. & De Luca, H.F. (1981) *Biochemistry
20*, 222–226.

23. Hains, A.H. (1976) *Adv. Carbohydr. Chem. Biochem. 33*, 11–109.

24. Tsuda, Y. & Yoshimoto, K. (1981) *Carbohydr. Res. 87*, C1–C4.

25. Sinclair, K.A. & Caldwell, J. (1981) *Biochem. Soc. Trans. 9*, 215.

26. Ljevaković, Dj. & Keglević, D. (1980) *Carbohydr. Res. 86*, 43–57.

27. Illing, P.A. & Wilson, I.D. (1981) *Biochem. Pharmacol. 30*, 3381–
3384.

28. Eggers, N.J. & Doust, K. (1981) *J. Pharm. Pharmacol. 33*, 123–
124.

29. Hignite, C.E., Tachanz, C., Lemons, S., Wiese, H., Azarnoff, L. &
Huffman, D.H. (1981) *Life Sci. 28*, 2077–2081.

30. Nemoto, N. & Gelboin, H. (1975) *Arch. Biochem. Biophys.   170*,
739–742.

31. Chasseaud, L.F., Down, W.H., Grover, P.L., Sacharin, R.M. & Sims,
P. (1980) *Biochem. Pharmacol. 29*, 1589–1590.

32. Delbressine, L.P.C., Seutter-Berlage, F. & Seutter, E.   (1981)
Xenobiotica 11, 241–247.

33. Van Bladeren, P.J., Delbressine, L.P.C., Hoogeterp, J.J., Beaumont,
A.H.G.M., Breimer, D.D., Seutter-Berlage, F. & van der Gen, A.
(1981) *Drug Metab. Dispos. 9*, 246–249.

34. van Bladeren, P.J., Breimer, D.D., Seghers, C.J.R., Vermeulen,
N.P.E., van der Gen, A. & Cauvet, J. (1981) *Drug Metab. Dispos.
9*, 207–211.

35. Seutter-Berlage, F., Delbressine, L.P.C., Smeets, F. L. M.   &
Ketalaars, H.C.J. (1978) *Xenobiotica 8*, 413–418.

36. Watabe, T., Hiratsuka, A., Ozawa, N. & Isobe, M. (1981) *Biochem.
Pharmacol. 30*, 390–392.

37. Plummer, J.L., Smith, B.R., Ball, L.M. & Bend, J.R. (1980) *Drug
Metab. Dispos. 8*, 68–72.

38. Mangold, J.B. & Abdel-Monem, M.M. (1980) *Biochem. Biophys. Res. Comm. 96*, 333-340.
39. van Doorn, R., Bos, R.P., Brouns, R.M.E., Leijdekkers, Ch.-M. & Henderson, P. Th. (1980) *Arch. Toxicol. 43*, 293-304.
40. Burstein, S. & Lieberman, S. (1958) *J. Biol. Chem. 233*, 331-335.
41. Goodall, A.B. & James, V.H.T. (1981) *J. Steroid Biochem. 14*, 465-471.
42. Baldwin, M.K. & Hutson, D.H. (1980) *Xenobiotica 10*, 135-144.
43. Islam, M.A., Raicht, R.F. & Cohen, I.B. (1981) *Anal. Biochem. 112*, 371-377.
44. Bell, J.A., Bradbury, A., Martin, L.E. & Tanner, R.J.N. (1981) *Xenobiotica 11*, 841-847.
45. Crayford, J.V. & Hutson, D.H. (1980) *Xenobiotica 10*, 355-364.
46. Tateishi, M., Koitabashi, C. & Ichihara, S. (1980) *Biochem. Pharmacol. 29*, 2705-2708.
47. Kirkpatrick, R.B., Lack, L. & Killienberg, P.G. (1980) *J. Biol. Chem. 255*, 10157-10159.
48. Krstulovic, A.M., Bertani-Dziedzic, L., Dziedzic, S.W. & Gitlow, S.E. (1981) *J. Chromatog. 223*, 305-314.
49. von Loffler, S. & Bolt, H.M. (1980) *Arzneim.-Forsch. 30*, 810-813.
50. Barford, P.A., Olavesen, A.H., Curtis, C.G. & Powell, G.M. (1977) *Biochem. J. 168*, 373-377.
51. Samarajeewa, P., Leepipatpaiboon, S. & Coulson, W.F. (1980) *Steroids 36*, 611-618.
52. Gottschlich, R. & Meltzer, M. (1980) *Xenobiotica 10*, 317-327.
53. Almé, B. & Sjövall, J. (1980) *J. Steroid Biochem. 13*, 907-916.
54. Mattox, V.R. & Litwiller, R.D. (1980) *J. Chromatog. 189*, 33-42.
55. Matsui, M. & Aoyagi, S. (1979) *Biochem. Pharmacol. 28*, 1023-1028.
56. Slikker, W., Jr., Lipe, G.W. & Newport, G.D. (1981) *J. Chromatog. 224*, 205-219.
57. Weitering, J.G., Krugsheld, K.R. & Mulder, G.J. (1979) *Biochem. Pharmacol. 28*, 757-762.
58. Chen, K.-C. & Dorough, H.W. (1979) *Drug Chem. Toxicol. 2*, 331-354.
59. Pang, K.S. & Terrell, J.A. (1981) *Biochem. Pharmacol. 30*, 1959-1965.
60. Filer, C.W., Humphrey, H.J., Jeffery, D.J., Jones, K.H. & Langley, P.F. (1980) *Xenobiotica 10*, 761-769.
61. Shirkey, R.J., Kao, J., Fry, J.R. & Bridges, J.W. (1979) *Biochem. Pharmacol. 28*, 1461-1466,
62. Hall, B.E. & James, S.P. (1980) *Xenobiotica 10*, 421-434.
63. Autrup, H. (1979) *Biochem. Pharmacol. 28*, 1727-1730.
64. Schenk, J. & Neuman, H.-G. (1980) *Xenobiotica 10*, 675-688.
65. Szinicz, L.L. & Weger, N. (1980) *Xenobiotica 10*, 611-620.
66. Elbers, R., Kampffmeyer, H.G. & Rabes, H. (1980) *Xenobiotica 10*, 621-632.
67. Son, O.S., Everett, D.W. & Fiala, E.S. (1980) *Xenobiotica 10*, 457-468.

68. Jones, D.P., Sundby, G.-B., Ormstad, K. & Orrenius, S.   (1979)
    Biochem. Pharmacol. 28, 929–935.
69. Grafström, R., Ormstad, K., Moldéus, P. & Orrenius, S.   (1979)
    Biochem. Pharmacol. 28, 3573–3579.
70. von Siegers, C.-P., Younes, M. & Oltmanns, D. (1980)   *Arzneim.-
    Forsch./Drug Res. 30*, 804–807.
71. Sanborn, H.R. & Malins, D.C. (1980) *Xenobiotica 10*, 193–200.

#C-2

ISOLATION AND CHARACTERIZATION OF AMINO ACID AND SUGAR CONJUGATES OF XENOBIOTIC CARBOXYLIC ACIDS

John Caldwell, Andrew J. Hutt,
Mary Varwell Marsh and Keith A. Sinclair

Department of Pharmacology
St. Mary's Hospital Medical School
London W2 1PG, U.K.

The carboxyl group in xenobiotics such as certain drugs and pesticides may be conjugated with glucuronic acid, glycine, glutamine or taurine. A logical system, developed for excreta but of general applicability, is now presented for isolating and characterizing such conjugates. The choice of TLC or HPLC system is determined by the mobility of the parent acid. It cannot be taken for granted that all alkali-labile conjugates are ester glucuronides. Problems in splitting off and identifying different conjugate moieties are considered. Chemical synthesis of amino acid conjugates is facile and aids identification.

Carboxylic acids are compounds of considerable economic importance, finding application as drugs, herbicides and insecticides among other uses. Since the human population may be deliberately or accidentally exposed to such compounds, their safety must be evaluated, with cardinal help from metabolic studies. Carboxylic acids may undergo diverse metabolic reactions in the body, either at the carboxyl group or elsewhere in the molecule. The numerous biotransformation options open to the carboxyl group [1] include decarboxylation, reduction, conversion into methyl esters or amides, and conjugation with sugars, notably glucuronic acid, or amino acids. Of these possibilities, glucuronic acid and amino acid conjugations are notably important [1,2], and their relative extents are a function of the structure of the acid and the animal species in question [2-4].

The importance of conjugation reactions in drug metabolism is frequently under-emphasized, partly because of difficulties in the analysis of such metabolites, which are generally highly polar and water-soluble. Having long been engaged in studying the conjugation

carboxylic acids of various types, we now describe generally applicable methodology developed recently for the isolation and characterization of the various conjugates of arylacetic acids in excreta.

As must be appreciated, a given acid may be conjugated with
glucuronic acid and with amino acids, so that the urine can contain
3 or 4 conjugates besides the free acid and other metabolites in
which other regions of the molecule have been transformed. The
analytical approaches to be described are thus intended to provide
general methods for the characterization of conjugates of carboxylic
acids present in mixtures.

PRELIMINARY INVESTIGATIONS

Before the choice of specific methods for characterizing amino
acid and sugar conjugates of arylacetic acids is considered, the
types of conjugate present must be assessed. This may be achieved
by subjecting urine, from animals or volunteers given the compound,
to TLC and/or HPLC in the systems to be described later, before and
after treatment with mild alkali (1 vol. M NaOH at 37° for 30 min).
Ester glycosides are broken down under these conditions, but amino
acid conjugates ('peptides') are untouched. At this stage the major
requirement is for a system in which the acid has a high R_f value (or
long R_t) and in which the more polar conjugates will have lower R_f
values (or short R_t values, provided that they are not eluted in the
void volume). Chromatograms showing the principles are depicted in
Fig. 1.

Having established the relative proportions of alkali-stable and
alkali-labile conjugates present, one can then proceed to the characterization of these two classes of metabolites separately.

Fig. 1. Stylized chromatograms
showing the preliminary step in
the characterization of the conjugates of a xenobiotic acid.
Trace **A** shows untreated urine,
and trace **B** the same urine after
treatment with M NaOH.

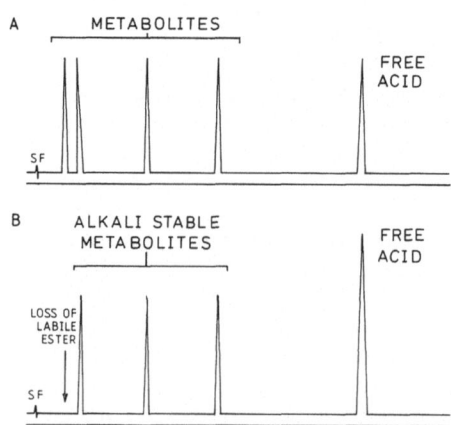

ALKALI-LABILE CONJUGATES

Ester glycosides, viz. conjugates of the xenobiotic acid with various sugars or sugar acids, are commonly encountered as metabolites. The commonest ester glycosides are, of course, glucuronides; but conjugates with glucose, ribose and xylose may be encountered [1, 5].

Although it is often assumed that all alkali-labile conjugates of acids are ester glucuronides, this is not justified and the analyst must take steps towards a positive identification of the conjugating agent. Xenobiotic glucuronides are commonly identified by their lability towards β-glucuronidase, which is available commercially at low cost from mammalian, molluscan and bacterial sources. However, it is important to remember that (a) the use of this enzyme requires both positive and negative control incubations, and (b) the chemistry of the ester glucuronides may render them resistant to the enzyme.

The susceptibility of a conjugate to β-glucuronidase may only be taken as evidence that the conjugate is indeed a β-D-glucopyranosiduronate when the following conditions have been met:
1: that the cleavage is inhibited by saccharo-1,4-lactone;
2: that boiled enzyme is inactive;
3: that the enzyme cleaves an authentic glucuronide, such as phenolphthalein glucuronide, when put into the biological sample;
4: that phenolphthalein glucuronide is not cleaved when saccharo-1,4-lactone is included.

Saccharo-1,4-lactone is a specific inhibitor of β-glucuronidase [6], and the controls 2 and 3 show that there is no spontaneous hydrolysis under the incubation conditions and that there are no inhibitors of the enzyme in the biological sample. Control 4 is included to show that the cleavage of the authentic glucuronide is indeed due to β-glucuronidase.

These conditions may be met readily for ether-type glucuronides but certain problems present themselves in the case of ester glucuronides. It is a common finding that the yield of aglycone from a putative ester glucuronide upon β-glucuronidase hydrolysis is less than that obtained by alkaline hydrolysis [refs. in 7, 8]. When this occurs, it is important to obtain evidence as to the nature of the conjugating agent present and to consider possible reasons for the failure of β-glucuronidase hydrolysis.

IDENTIFICATION OF THE SUGAR-CONJUGATING AGENT

A wide variety of tests are available for this purpose, but many of them require the availability of the conjugate in a reasonably pure form [9, 10]. We have found that ammonolysis on TLC plates is the quickest and most convenient method [7, 11]. The urine is first

subjected to preparative TLC, and the band corresponding to the conjugate under study is scraped off the plate and eluted with methanol.
The concentrated eluate is then applied to the TLC plates, which are
placed overnight in a tank containing a small beaker of 0.88 s.g.
ammonia.   The plates are removed, placed in a fume hood to remove
traces of ammonia, and then standards of possible sugar-conjugating
agents, such as glucuronic acid, glucose, xylose and ribose, are
spotted alongside the sample. The plates are developed in a suitable
solvent (we have found propanol/water, 17:3, very successful on
silica gel plates) and, after drying, sprayed with naphthoresorcinol.
The sugars are seen as blue spots, and comparison of R_f values permits identification of the unknown.   Other plates may be spotted
with the aglycone and developed in appropriate systems, to show
whether the aglycone is the parent acid or a phase I metabolite.

This method is sufficiently sensitive to detect 0.5 μmol amounts
of the sugars.

WHY ARE SOME ESTER GLUCURONIDES RESISTANT TO β-GLUCURONIDASE ?

It has been known since 1920 that many substituted sugars may
undergo intramolecular rearrangement of the aglycone, in which the
aglycone moves from -OH to an adjacent -OH [12]. This is especially
well known for acyl constituents, and although carbohydrate chemists
are familiar with such migrations [13], drug metabolists have largely
ignored the phenomenon. β-Glucuronidase is specific for 1-O-substi-
tuted β-D-glucopyranosiduronates, and other stereo- and geometric
isomers are not hydrolysed [14].

In 1978, Heirwegh, Compernolle and colleagues [15, 16] demonstrated unequivocally that bilirubin IXα monoglucuronide underwent
facile intramolecular rearrangement at mildly alkaline pH, leading
to the formation of the 2-, 3- and 4-O-substituted esters of glucuronic acid from the biosynthetic 1-O-acyl glucuronide. Following
this work, Faed [8] suggested that the same phenomenon occurred with
the ester glucuronide of clofibric acid (p-chlorophenoxyisobutyric
acid), a widely used hypolipidaemic agent.   She found two alkali-
labile conjugates in human urine, only one of which was acted on
by β-glucuronidase, and postulated that they were isomers.

As part of a study on the metabolism of clofibric acid in a
range of mammalian species [17, 18], a fully authenticated sample of
1-O-clofibryl β-D-glucopyranosiduronate was obtained, by isolation
from rabbit urine [19].   A study of the possible intramolecular rearrangement of xenobiotic ester glucuronides was therefore initiated
with this compound.

Solutions of clofibrylglucuronide were incubated in buffers
from pH 5.2 to 8.6 for 3 h at 37°, and these were then (a) assayed for
aglycone released during the pre-incubation, (b) incubated at pH 5

with β-glucuronidase (with and without 2 mg saccharo-1,4-lactone) followed by assay of aglycone, and (c) heated to 37° for 4 h with an equal volume of M NaOH, followed by assay of the aglycone.

The % of the total clofibric acid present that was released by β-glucuronidase was highly dependent upon the pH of pre-incubation. The values found by Sinclair & Caldwell [7] for % resistant to the enzyme were nil for pH 5.2, 6.0 or 7.0, but were 8, 34 and 84% respectively for pH 7.4, 8.0 and 8.6, i.e. with pH 8.6 there was only 16% cleavage. HPLC analysis showed that the initial solutions contained a single compound, which was susceptible to β-glucuronidase, but upon incubation at alkaline pH, 3 new peaks were present as well as the original compound and the free acid. None of these new peaks were affected by β-glucuronidase treatment, and all the compounds responsible for them gave a positive naphthoresorcinol reaction, showing the presence of glucuronic acid and indicating that they are clofibric acid esters of glucuronic acid not involving the hydroxyl group at the anomeric carbon (C-1) of glucuronic acid. Thus far, the 1-O-acyl glucuronide has been characterized by proton and ^{13}C-NMR spectroscopy and CI-MS, and the other isomers are currently being investigated. When the aglycone moves away from C-1, mutarotation is possible, as are the interconversion of pyranose, furanose and open-chain forms and the formation of 3,6-lactones. The number of possible structures is thus considerable.

This intramolecular rearrangement of ester glucuronides from the biosynthetic 1-isomers to β-glucuronidase-resistant forms has implications for the use of this enzyme to characterize conjugates of carboxylic acids. Similar results to the above have been obtained with the urine of animals and human volunteers dosed with clofibric acid, even when the urine is collected over solid CO_2. It would appear from a re-evaluation of previous work from this laboratory and the literature that the ester glucuronides of many carboxylic acids undergo, at least partially, rearrangement to β-glucuronidase-resistant forms [7; see also 20].-
Known for: bilirubin, clofibric acid, probenecid, isoxepac, and also for Wy-18251 (F.W. Janssen & H. Ruelius, pers. comm.) and fenclofenac (M.V. Marsh & J. Caldwell, unpublished).
Suspected for: benzoic acid, hydratropic acid, 1-naphthylacetic acid, 2-naphthylacetic acid, indol-3-ylacetic acid, diphenylacetic acid, fenoprofen, indomethacin, buniodyl and tyropanoic acid.

ANALYTICAL METHODOLOGY FOR XENOBIOTIC ESTER GLUCURONIDES

If the results obtained by β-glucuronidase treatment, properly controlled as outlined above, agree with those obtained by alkaline hydrolysis, there is excellent evidence for the presence of a glucuronic acid conjugate. However, if these experiments give different results, it is necessary to investigate the metabolite further. The ammonolysis procedure outlined previously will permit the identification of the conjugating agent. If the conjugate is suspected to undergo intramolecular rearrangement, the process cannot be reversed, but the possibility can be studied by incubation at pH 8.6 followed by β-glucuronidase treatment. If the lability to β-glucuronidase is reduced by this treatment, then it gives evidence that intramolecular migration is probably responsible for the original partial β-glucuronidase resistance.

It is possible to separate the various isomers by HPLC using conventional RP and ion pair-RP systems. In RP systems, reducing the proportion of organic solvent, while increasing the retention of aglycone, can result in separation of the various isomers, which generally fuse as a single highly polar peak. Thus, with a C-18 column and a mobile phase of 70% methanol containing 0.1% trifluoroacetic acid the various isomers of clofibryl glucuronide run as a single peak, but with 40% methanol the isomers are separated into 4 discrete peaks [20].

ANALYSIS OF ALKALI-STABLE CONJUGATES

Separation and isolation

If there are metabolites present in the body fluids which are resistant to mild alkaline treatment, but which are converted into the free acid or a phase I metabolite thereof by strong alkaline or acid treatment, it is likely that they are conjugates with amino acids. Although glycine, glutamine and taurine are most frequently encountered in xenobiotic conjugates, a variety of other amino acids have been found, including ornithine, alanine, serine, glycylglycine and glycyltaurine [1-4, 21]. A variety of methods are available for the isolation and characterization of amino acid conjugates, the three commonest types of which are the focus for this article.

TLC

TLC of suspected amino acid conjugates in the sample of interest (after alkaline treatment) in standard systems often provides important information towards their identification. From various studies in this laboratory, we have available authentic samples of the glycine, glutamine and taurine conjugates of phenyl, *p*-chlorophenyl, *p*-nitrophenyl-, α-methylphenyl- (hydratropic), 1- and 2-naphthyl-, indol-3-yl- and diphenyl- acetic acids.

Amongst various TLC systems investigated, that found to be most generally useful had a mobile phase of benzene/acetone/glacial acetic acid (6:2:1 by vol.) [22] with Merck 60-F_{254} silica gel plates (0.2 mm thick on aluminium support, cat. no. 5554; E. Merck, Darmstadt, W. Germany) developed to 15 cm from the origin.

Obviously, the R_F values of the various acids and their conju-

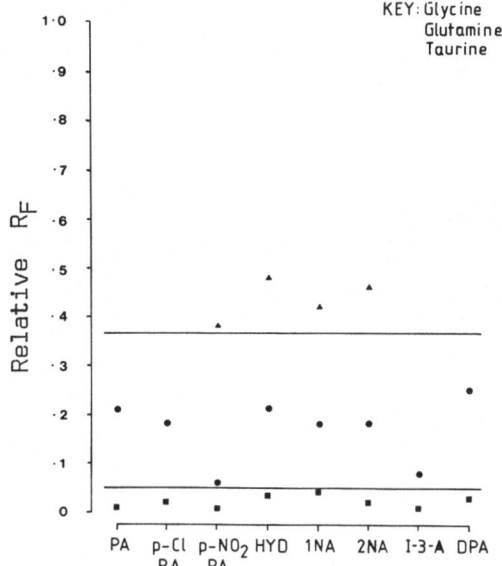

Fig. 2. Relative R_F values: various amino acid conjugates *vs.* their respective acids. TLC in benzene/acetone/glacial acetic acid, 6:2:1.

Key to aglycones:

PA, phenylacetic acid;
p-Cl PA, *p*-chlorophenylacetic acid;
p-NO$_2$PA, *p*-nitrophenylacetic acid;
HYD, hydratropic acid;
1NA, 1-naphthylacetic acid;
2NA, 2-naphthylacetic acid;
I-3-A, indol-3-ylacetic acid;
DPA, diphenylacetic acid.

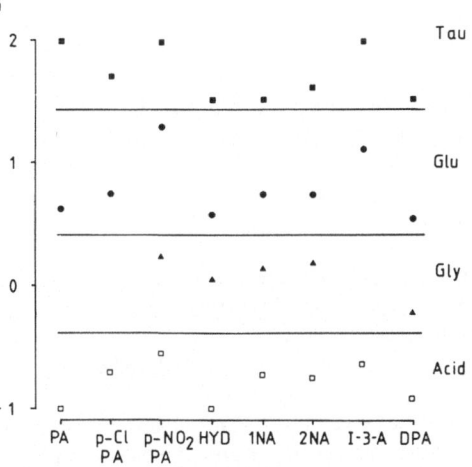

Fig. 3. R_m values (see text) for each acid and its glycine, glutamine and taurine conjugates.

TLC system and abbreviations as above.

gates are highly variable; but the relative R_F values for conjugate *vs.* parent acid provide a clear distinction between the different types of conjugate. These relative values (Fig. 2) show that conjugates fall into a well-defined order of polarity:

taurine > glutamine > glycine > parent acid.

Fig. 3 shows an alternative, and perhaps superior, comparison where R_m is calculated:

$$R_m = \log_{10}\left(\frac{1}{R_F} - 1\right)$$

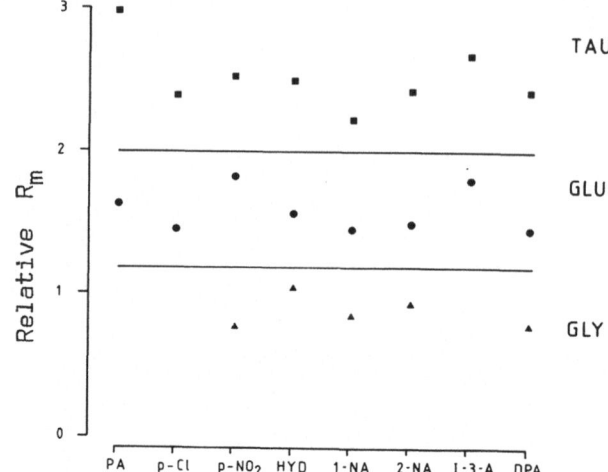

Fig. 4. Relative R_m values: conjugate R_m vs. parent acid R_m.

TLC system and abbreviations as in Fig. 2.

In this presentation, the various compounds fall into 4 distinct groups and a conjugate could be provisionally identified from its R_m value. The distinction between the types of conjugate is accentuated further if relative R_m values are calculated:

$$\text{Relative } R_m = \frac{R_m \text{ of conjugate}}{R_m \text{ of parent acid}}$$

These values are shown in Fig. 4, and it is clear that the relative R_m values for taurine conjugates are > 2, for glutamine conjugates 1.2 – 2.0, and for glycine conjugates 0.7 – 1.2.

HPLC

The series of arylacetic acids and their conjugates listed above was also examined by HPLC in a standard system. Fig. 5 shows the relative R_t values:*

$$\text{Relative } R_t = \frac{R_t \text{ of conjugate}}{R_t \text{ of parent acid}}$$

It is clear that in this presentation, while acceptable separations of the various conjugates are achieved, any predictive value is weakened by the two glutamine conjugates overlapping in relative R_t with the majority of glycine conjugates, and one overlapping with the taurine conjugates.

A superior presentation of these data uses the Kappa value, k', which takes account of the void volume of the system:

$$k' = \frac{\text{Retention volume – void volume}}{\text{void volume}}$$

Hence (Fig. 6): Relative k' value $= \dfrac{k' \text{ for conjugate}}{k' \text{ for parent acid}}$

_____
* $\equiv t_R$ as in HPLC texts.- *Ed.*

Fig. 5. HPLC R_t values for various amino acid conjugates relative to R_t for the parent acid.
Abbreviations: see Fig. 2 legend.
Waters U6K injector, M6000A pumps, M720 systems controller; Cecil 2012A detector.
Column: 100 X 5 mm 5 μm ODS-Hypersil. Mobile phase: 30% (v/v) methanol, adjusted with glacial acetic acid to pH 3.0; 2 ml/min. UV wavelength settings appropriate for each acid.

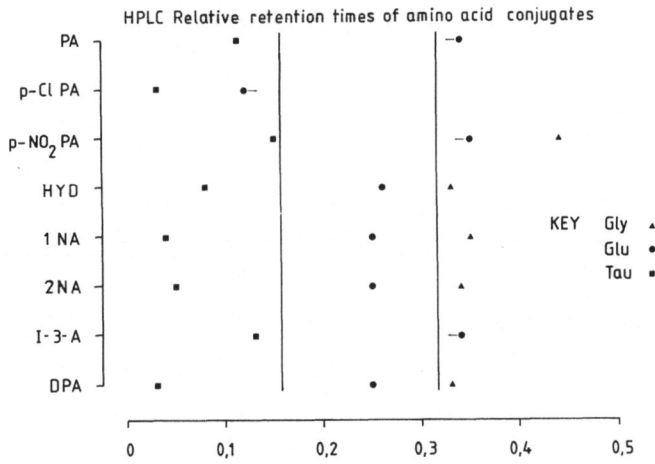

Fig. 6. Relative k' values: amino acid conjugates vs. their respective acids.
Abbreviations as for Fig. 2; HPLC as above.

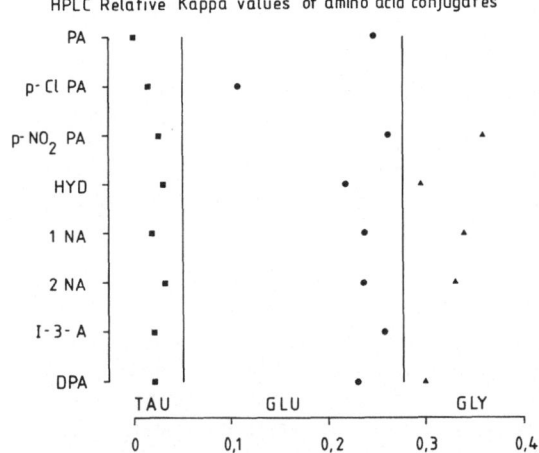

There is a definite order of elution, in which the taurine conjugates are retained least, followed by glutamine and glycine conjugates and the parent acids. Again, calculation of the relative k' value will give a provisional identification of an unknown conjugate.

Although the standard system with a mobile phase of pH 3.0 30% methanol is important as a starting point in such work, with more lipophilic acids the retention times may be too long for convenient analysis. If this is encountered, it is tempting to raise the concentration of methanol in the mobile phase; but this often reduces the length of the assay at the expense of the separation of the glycine and glutamine conjugates. It is better in such cases to use solvent programming.

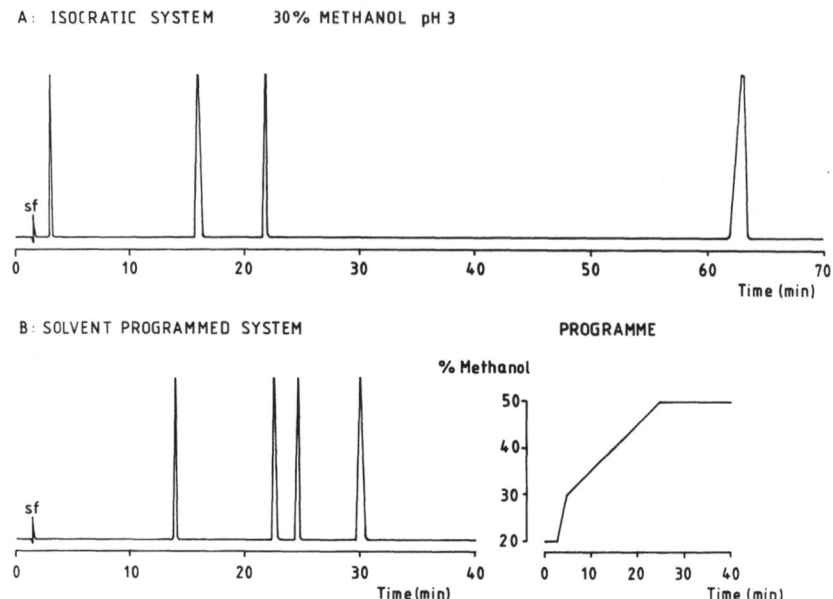

Fig. 7. Utility of solvent programming in the HPLC separation of amino acid conjugates. Illustrated for 2-naphthylacetic acid. The peaks correspond, in order of elution, to the taurine conjugate, glutamine conjugate, glycine conjugate and the free acid. **A**, the isocratic system; **B**, the programmed system (as in bottom-right graph). For details see Fig. 5 legend and text.

This is illustrated in Fig. 7 for 2-naphthylacetic acid and its amino acid conjugates. In the standard isocratic system the acid has R_t > 60 min. R_ts should ideally be as short as possible, both to reduce assay time and thereby increase sample throughput and save solvent, and to improve peak shape, which is particularly pertinent when an integrator is not used and compounds are quantified on the basis of peak height.

In this example, the use of solvent programming to alter the solvent composition during the run has halved the assay time. Pump A contained 20% methanol pH 3.0, and pump B 50% methanol pH 3.0; the gradient shown was achieved by use of the microprocessor controller. Evidently the peak shape is improved and baseline separation of the glycine and glutamine conjugates maintained. The taurine conjugate is retained upon the column, which is important for isolation and subsequent analysis since many endogenous urinary compounds (generally UV-absorbing) elute from the column in the void volume.

Thus the two chromatography systems, TLC and HPLC, show how a putative amino acid conjugate behaves relative to its parent acid, giving a good indication of its nature, and provide sufficient separation to allow isolation for subsequent identification by instrumental approaches such as NMR and MS.

NUCLEAR MAGNETIC RESONANCE SPECTROSCOPY (NMR)

One of the major analytical problems presented by the amino acid conjugates of carboxylic acids is that of their lack of volatility, which renders vapour phase methods difficult. NMR is able to give a great deal of structural information on the molecule while it is in solution, and with the increasing availability of high field (>200 MHz) spectrometers the drawback of the large sample size required is largely overcome.

We have obtained proton NMR spectra for synthetic glycine, glutamine and taurine conjugates of a variety of aromatic, arylacetic and aryloxyacetic acids, using a 250 MHz instrument. The sample size was 1-10 mg. The NMR spectra obtained are highly characteristic of the various amino acids, differing only in the signals from the various aromatic and alkylaromatic systems in the remainder of the molecule. The spectra shown in Figs. 8-10 are of the glycine, glutamine and taurine conjugates of 2-naphthylacetic acid, and serve to illustrate the principles obtaining for these types of conjugate.

NMR spectra of glycine conjugates are identical with those of the parent acids, with the addition of a signal at about δ 4.0 ppm equivalent to two protons. The multiplicity of this signal depends upon the solvent: thus, if the amide proton does not exchange with solvent (e.g. in $CDCl_3$) the signal will be a doublet, which converts to a singlet upon addition of D_2O or another protic solvent, or with spin-decoupling. Fig. 8 shows the NMR spectrum of 2-naphthylacetylglycine.

Fig. 9 gives the spectrum of 2-naphthylacetylglutamine, and is typical of the spectra of other glutamine conjugates. The glutamine moiety gives a characteristic double doublet at about δ 4.4 ppm, arising from the proton on the chiral centre (due to the non-equivalence of the adjacent methylene protons) and a characteristic complex multiplet between 1.9 and 2.4 ppm (equivalent to 4 protons) due to the protons upon the two methylene groups. The inset in Fig. 9 shows the spectrum of glutamine and a comparison of the two shows the downfield shifts of the signals from the proton on the chiral centre and from the adjacent methylene group upon formation of the peptide bond.

The spectra of taurine conjugates are similar to those of the parent acid, with the addition of a pair of 2-proton triplets at δ 3.5 and 3.0 ppm. The spectrum of 2-naphthylacetyltaurine in Fig. 10, obtained in DMSO-d_6, shows also the exchangeable amide proton at

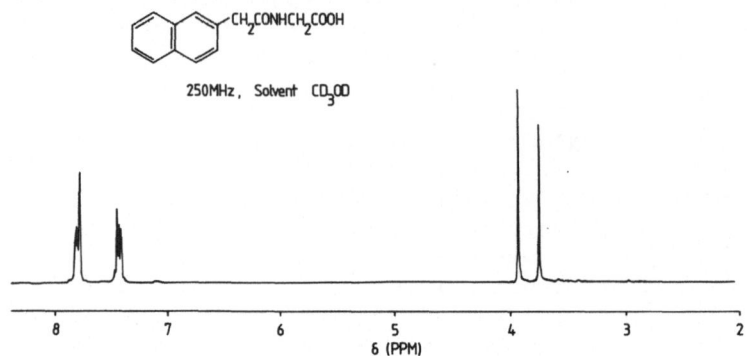

Fig. 8. NMR spectrum of the glycine conjugate of 2-naphthylacetic acid (250 MHz in CD₃OD; signals due to the solvent have been omitted). Signal at δ 3.75 ppm is due to the methylene protons of the glycine residue.

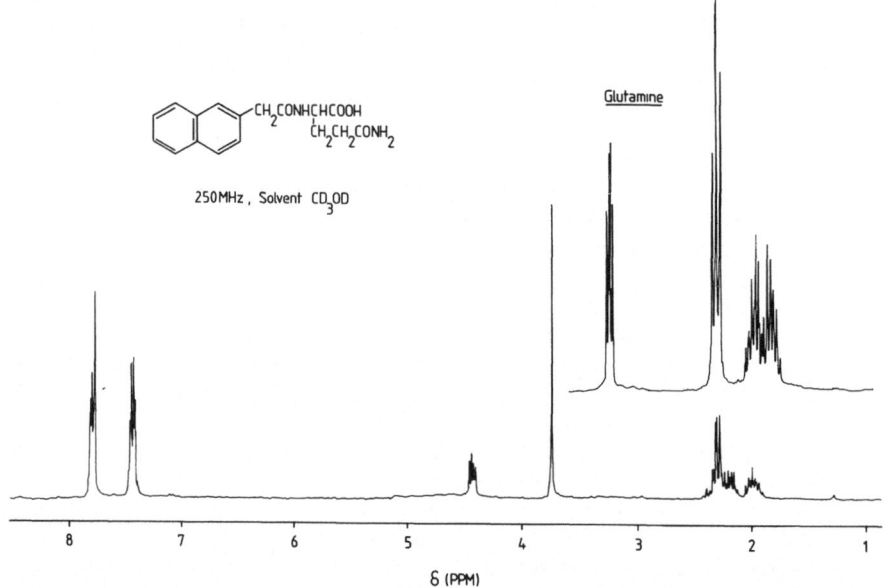

Fig. 9. NMR spectrum (as in Fig. 8) of the glutamine conjugate of 2-naphthylacetic acid. Signal at δ 4.42 ppm (double doublet) is due the methine proton of the glutamine residue. *Inset:* NMR spectrum of glutamine obtained under similar conditions.

about δ 8 ppm and the signal at δ 3.5 ppm as a multipet. The addition of D_2O to the DMSO-d_6 results in the loss of the signal at δ 8 ppm, and that at δ 3.5 ppm becomes a triplet (Fig. 10). This shows that the signal at δ 3.5 ppm arises from the methylene group

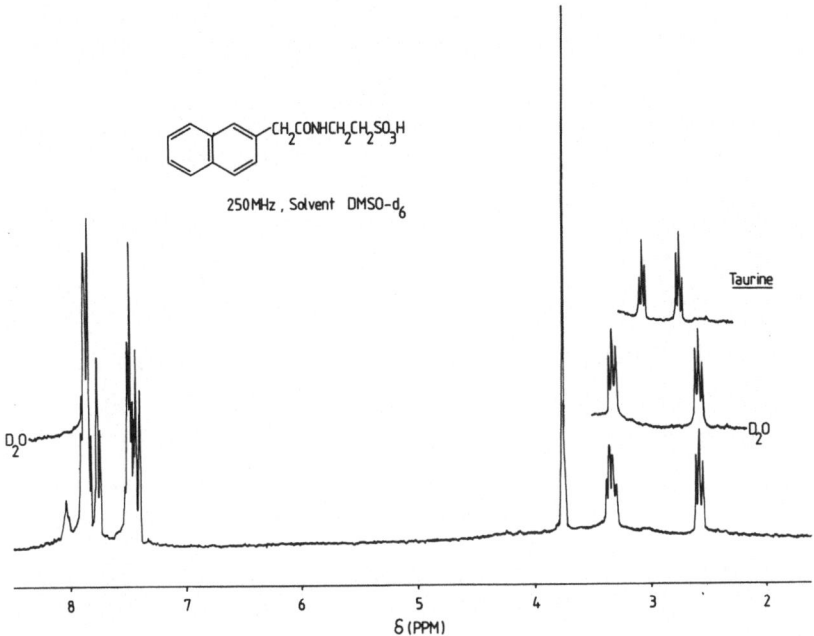

Fig. 10. NMR spectrum (as in Fig. 8, but in DMSO-d$_6$) of the taurine conjugate of 2-naphthylacetic acid, before and after addition
of D$_2$O. Signals at δ 3.56 and 3.06 are due to the methylene group
protons of the taurine moiety. *Inset:* NMR spectrum of taurine in
DMSO-d$_6$ plus D$_2$O.

adjacent to the peptide bond. Comparison of these spectra with that
of taurine (obtained in D$_2$O; *inset*) shows that when the conjugate
is formed, the signals of the –CH$_2$ adjacent to the amino group shift
downfield, and that those from the –CH$_2$ adjacent to the –SO$_3$H group
shift upfield.

MASS SPECTROMETRY (MS)

Since the amino acid conjugates are highly ionized and of low
volatility, their mass spectra are generally recorded after methylation with diazomethane. This increases their volatility and may
make them amenable to GC–MS [11]. Treatment of glycine conjugates
with diazomethane may produce mono- and di-methylated derivatives,
in which the carboxyl group is esterified and the amido nitrogen
methylated.

Tables 1–3 give characteristic ion fragments and their relative abundances for the fragmentation of methyl esters of glycine,
glutamine and taurine conjugates of various xenobiotic acids. All
spectra were obtained by electron impact (EI) at 70 eV. Although

Table 1. Characteristic ion fragments in the mass spectra of the methyl esters of glycine conjugates, based on examination of 20 compounds.

General formula $R.CONH.CH_2.COOCH_3$

Fragment ion	m/z	Relative abundance, %
$M^{+\cdot}$	M	0.1–89
$M^{+\cdot} - \dot{O}CH_3$	M – 31	2–7
$M^{+\cdot} - \dot{C}OOCH_3$	M – 59	4–16
$M^{+\cdot} - \dot{N}HCO_2.COOCH_3$	M – 88	1–100
$M^{+\cdot} - \dot{C}ONH.CH_2.COOCH_3$	M – 116	11–100
$[CH_3OCO.CH_2.NHCO]^+$	116	30–75
$[CH_3OCO.CH_2NH]^+$	88	40–100

Table 2. Characteristic ion fragments in the mass spectra of the methyl esters of glutamine conjugates, based on examination of 6 compounds.

General formula $R.CONH.CH.CH_2.CH_2.CONH_2$
                                          $COOCH_3$

Fragment ion	m/z	Relative abundance, %
$M^{+\cdot}$	M	0.5–10
$M^{+\cdot} - NH_2CH(COOCH_3).CH_2.CH_2.CONH_2$	M – 160	38–78
$M^{+\cdot} - CONH.CH(COOCH_3).CH_2.CH_2.CONH_2$	M – 187	47–100
$[CONH.CH(COOCH_3).CH_2.CH_2.CONH_2]^+$	187	54–100
$[\text{ same } (m/z\ 187) - CH_3OH]^+$	155	17–98
$[CH_3OCO.CH_2.NHCO]^+$	116	29–96

the range of relative abundances of the various ions is large, general fragmentation pathways may be discerned. Thus, in all spectra so far reported, there are fragments due to elimination of the amino acid moiety followed by loss of CO, which give intense ions. Fragments arising from the amino acid residue are also seen.

Of the three types of amino acid conjugate examined, mass spectra are obtainable for glycine conjugates by direct insertion or by GC–MS of the methyl esters. Glutamine conjugates as their methyl esters are best investigated by direct insertion. Taurine conjugates present the greatest difficulty as (a) they are frequently isolated

Table 3. Characteristic ion fragments in the mass spectra of the methyl esters of taurine conjugates, based on examination of 9 compounds.

General formula $R.CONH.CH_2.CH_2.SO_3CH_3$

Fragment ion	m/z	Relative abundance, %
$M^{+\cdot}$	M	0–47
$M^{+\cdot} - CH_2.CH_2.SO_3CH_3$	M – 123	2–44
$M^{+\cdot} - NH.CH_2.CH_2.SO_3CH_3$	M – 138	1–100
$M^{+\cdot} - CONH.CH_2.CH_2.SO_3CH_3$	M – 166	11–100
$[CONH.CH_2.CH_2.SO_3CH_3]^+$	166	6–90
$[CONH.CH_2.CH_2.SO_2]^+$	134	7–80

as mixtures of the free acids and various salts, which renders= their derivatization difficult, and (b) they tend to decompose upon the MS probe or during GC–MS, giving several products of unknown structure. All the spectra reported in Table 3 were obtained by the direct insertion method.

The above discussion makes clear the need for further investigations of alternative derivatization procedures and ionization modes which may prove superior to currently used techniques.

SYNTHESIS OF AMINO ACID CONJUGATES

As a final step in characterizing amino acid conjugates, it is desirable to compare the chromatographic and spectral properties of the putative metabolite with those of a fully authenticated synthetic sample. The availability of a number of facile syntheses for amino acid conjugates renders this final step in their characterization comparatively easy.

Although the first synthesis of hippuric acid reported in the literature by Dessaignes [23] involved the fusion of benzoic acid and glycine in a sealed tube at 160° for 12 h, modern methods are less drastic. Generally, the carboxyl group of the xenobiotic is activated prior to reaction with the nucleophilic amino group of the amino acid, via an addition–elimination reaction, giving the peptide bond.

(a) Schotten–Baumann reaction

The xenobiotic acid chloride is prepared by refluxing the acid with an excess of thionyl chloride in toluene. The acyl chloride is then reacted with the amino acid in dilute alkaline solution, which both catalyses the reaction and neutralizes the HCl produced. This reaction is probably the most widely used of all procedures for the synthesis of amino acid conjugates.

$$Ar.COOH \xrightarrow{\;SOCl_2\;} Ar.COCl \tag{1}$$

$$Ar.COCl + H_2NR.COOH \xrightarrow{\;NaOH\;} Ar.CONHR.COOH + NaCl + H_2O \tag{2}$$

Full accounts of the use of the above reaction sequence in the synthesis of glycine, glutamine and taurine conjugates of several acids are given by Dixon et al. [24-26].

The Schotten–Baumann reaction may be unsuccessful when applied to syntheses of amino acid conjugates of acids having a functional group, e.g. –OH or –NH$_2$, which may react with the acid chloride, giving rise to homoconjugates. In such cases, alternative reaction sequences must be employed, such as the following two.

(b) Mixed anhydride reactions

In these syntheses the acid is activated by the formation of a mixed anhydride with a haloacid ester, e.g. ethyl chloroformate, in the presence of base:

$$Ar.COO^- + Cl.COOC_2H_5 \longrightarrow Ar.CO.O.COOC_2H_5$$

or with *N*-ethylcarbonyl-2-ethoxy-1,2-dihydroquinoline:

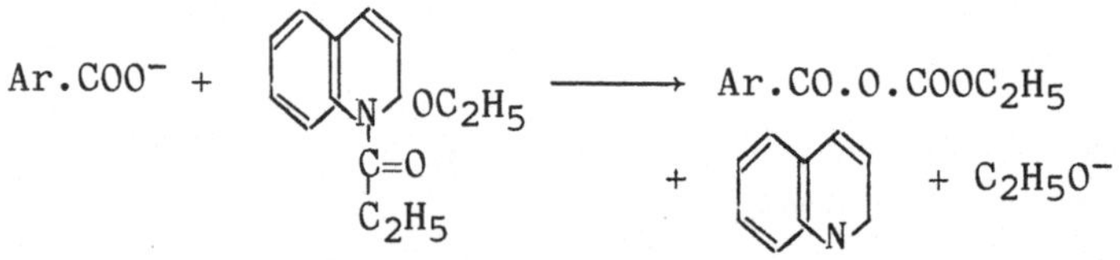

The mixed anhydride then reacts with the amino group of the amino acid as follows:

$$Ar.CO.O.COOC_2H_5 + H_2N.R'.COOH \longrightarrow Ar.CONH.R'.COOH$$
$$+ CO_2\!\uparrow + C_2H_5OH$$

A full method for the synthesis of indol-3-ylacetylglutamine via a mixed anhydride is given by Bridges et al. [27].

(c) Carbodiimide syntheses

The acid is activated by adduct formation with a carbodiimide, generally dicyclohexylcarbodiimide:

$$C_6H_{11}N{=}C{=}NC_6H_{11} + Ar.COO^- + H^+ \longrightarrow C_6H_{11}N{=}\underset{\underset{O.CO.Ar}{|}}{C}{-}NHC_6H_{11}$$

This adduct, in which the —COO—C≡N— function behaves as an anhydride, then reacts with the amino acid:

$$C_6H_{11}N=\underset{\underset{\displaystyle O.CO.Ar}{|}}{C}-NHC_6H_{11} \quad + \quad H_2N.R.COOH$$

$$\longrightarrow Ar.CO.NH.R.COOH + C_6H_{11}NHCONHC_6H_{11}$$

The disubstituted urea liberated in this reaction is very insoluble and so precipitates out of solution.

The above method has proved successful for the synthesis of a number of glycine conjugates [18, 28], but attempts to apply such procedures to the synthesis of taurine conjugates have been largely unsuccessful, since taurine is very poorly soluble in the solvents used (tetrahydrofuran, etc.). Recently, water-soluble carbodiimide reagents have become available, and these may prove satisfactory for the preparation of taurine conjugates.

CONCLUDING COMMENTS

Taken together, the various analytical approaches to the characterization of the amino acid conjugates of xenobiotic acids permit the relatively rapid and facile identification of such metabolites. It is important that the different techniques are applied in a systematic fashion, and Scheme 1 gives what our experience has shown to be a logical and fruitful way to proceed.

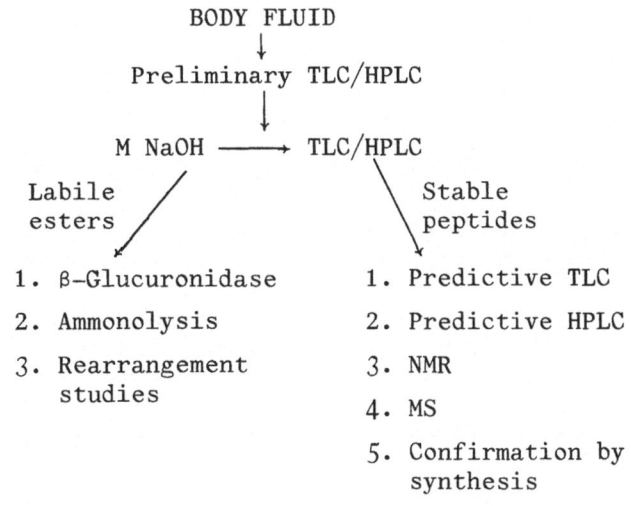

Scheme 1. Scheme for the systematic investigation of amino acid and sugar conjugates of xenobiotic acids in body fluids.

```
                        BODY FLUID
                            ↓
                 Preliminary TLC/HPLC
                            ↓
               M NaOH ——→ TLC/HPLC
                  Labile  /          \  Stable
                  esters /            \ peptides
                        ↙              ↘
            1. β-Glucuronidase      1. Predictive TLC
            2. Ammonolysis          2. Predictive HPLC
            3. Rearrangement        3. NMR
               studies             4. MS
                                    5. Confirmation by
                                       synthesis
```

For some compounds, other analytical techniques besides those mentioned here may be employed. The problems of dealing with those conjugates in which other parts of the parent xenobiotic acid have been metabolically transformed have had scant attention here, though obviously MS and NMR will aid in such cases. Chemical hydrolysis of amino acid conjugates and subsequent analysis of the hydrolysate by the methods of amino acid chemistry has also proved useful, notably for the taurine conjugates [8, 29, 30] which indeed were first dis- covered in this way [30].

In our laboratory we have used extensively the analytical meth- odology now described, in metabolic studies on a variety of xenobio- tic acids of diverse structure. We are sure that these approaches will prove of value to other workers interested in this important group of foreign compounds.

Acknowledgements

A.J. Hutt is a postdoctoral Fellow supported by the Sir Halley Stewart Trust. Mary Varwell Marsh and K.A. Sinclair are SRC–CASE students, in collaboration with Racecourse Security Services Labora- tories and ICI Pharmaceuticals respectively. HPLC equipment was pur- chased with grants from Cilag–Chemie Stiftung and the University of London Central Research Fund. The encouragement of Professor R.L. Smith is gratefully acknowledged.

References

1. Caldwell, J. (1982) in *Metabolic Basis of Detoxication* (Jakoby, W.B., Bend, J.R. & Caldwell, J., eds.), Academic Press, New York, 271–290.
2. Caldwell, J., Idle, J.R. & Smith, R.L. (1980) in *Extrahepatic Metabolism of Drugs and Other Foreign Compounds* (Gram, T.E., ed.), SP Medical & Scientific Books, New York, pp. 453–477.
3. Caldwell, J. (1978) in *Conjugation Reactions in Drug Biotrans- formation* (Aitio, A., ed.), Elsevier/N.Holland, Amsterdam, pp. 111–120.
4. Caldwell, J. (1980) in *Enzymatic Basis of Detoxication* (Jakoby, W.B., ed.), Academic Press, New York, pp. 85–114.
5. Heirwegh, K.P.M. (1978) in *Conjugation Reactions in Drug Bio- transformation* (Aitio, A., ed.), Elsevier/N. Holland, Amsterdam, pp. 67–76.
6. Levvy, G.A. & Conchie, J. (1966) in *Glucuronic Acid Free and Combined* (Dutton, G.J., ed.), Academic Press, New York, pp.301–364.
7. Sinclair, K.A. & Caldwell, J. (1982) *Biochem. Pharmacol. 31*, in press.
8. Faed, E.M. & McQueen, E.G. (1978) *Clin. Exp. Pharmacol.Physiol. 5*, 195–198.
9. Keglević, D. (1979) *Adv. Carbohyd. Chem. Biochem. 36*, 57–134.
10. Heirwegh, K.P.M. & Compernolle, F. (1979) *Biochem. Pharmacol. 28*, 2109–2114.
11. Marsh, M.V., Caldwell, J., Smith, R.L., Horner, M.W., Houghton, E. & Moss, M.S. (1981) *Xenobiotica 11*, 655–663.

12. Fischer, E. (1920) *Ber. 53,* 1621–1633.
13. Haines, A.H. (1976) *Adv.Carbohyd. Biochem. 33,* 11–109.
14. Marsh, C.A. (1966) *Glucuronic Acid Free and Combined* (Dutton, G.J., ed.), Academic Press, New York, pp. 27–136.
15. Compernolle, F., Van Hees, G.P., Blanckaert, N. & Heirwegh, K.P.M. (1978) *Biochem. J. 171,* 185–201.
16. Blanckaert, N., Compernolle, F., Leroy, P., Van Hautte, R., Fevery, J. & Heirwegh, K.P.M. (1978) *Biochem. J. 171,* 203–214.
17. Caldwell, J., Emudianughe, T.S. & Smith, R.L. (1979) *Brit. J. Pharmacol.* 66, 421P–422P.
18. Emudianughe, T.S., Caldwell, J. & Smith, R.L. (1982) *Drug Metab. Dispos.* in press.
19. Caldwell, J. & Emudianughe, T.S. (1979) *Biochem. Soc Trans.* 7, 521–522.
20. Sinclair, K.A. & Caldwell, J. (1981) *Biochem. Soc. Trans.* 9, 215.
21. Caldwell, J. (1982) *Drug Metab. Rev.* in press.
22. Idle, J.R., Millburn, P. & Williams, R.T. (1976) *Xenobiotica* 8, 253–264.
23. Dessaignes, V. (1857) *J. Pharm. (Paris)* 32, 44–47.
24. Dixon, P.A.F., Caldwell, J. & Smith, R.L. (1977) *Xenobiotica* 7, 695–706.
25. as for 24., 707–715.
26. as for 24., 717–725.
27. Bridges, J.W., Evans, M.E., Idle, J.R., Millburn, P., Osiyemi, F.O., Smith, R.L. & Williams, R.T. (1974) *Xenobiotica* 4, 645–652.
28. Solheim, E. & Scheline, R.R. (1973) *Xenobiotica* 3, 493–510.
29. Emudianughe, T.S., Caldwell, J., Dixon, P.A.F. & Smith, R.L. (1978) *Xenobiotica* 8, 525–534.
30. James, M.O., Smith, R.L., Williams, R.T. & Reidenborg, M.M. (1972) *Proc. Roy. Soc. Ser. B 182,* 25–35.

#C- 3

CHROMATOGRAPHY, HYDROLYSIS AND REARRANGEMENT OF THE GLUCURONIDE CONJUGATE OF ISOXEPAC, AS AFFECTING ITS ASSAY IN URINE

I.D. Wilson, A. Bhatti, H.P.A. Illing,
T.A. Bryce and J. Chamberlain

Hoechst Pharmaceutical Research Laboratories
Milton Keynes, Bucks. MK7 7AJ, U.K.

In man the non-steroidal anti-inflammatory drug isoxepac is excreted into the urine, mainly as an ester glucuronide. A pH-dependent conversion of the presumed 1-O-acyl glucuronide into several enzyme-resistant conjugates is observed. All the rearrangement products may be hydrolyzed, using mild alkali, to isoxepac. This conversion into β-glucuronidase-resistant conjugates has a bearing on the assay of isoxepac and its glucuronides, and of other drugs excreted as ester glucuronides. Suitable procedures for the determination of total isoxepac in human urine are described.

The ester glucuronide, presumably 1-O-acyl, which is the main urinary excretion product of isoxepac [6,11-dihydro-11-oxodibenz-(b,e)oxepin-2-acetic acid] in man [1] is readily hydrolysed under conditions that may occur on storage. Since much of the free drug detected in a sample may have arisen from hydrolysis, the distinction between the free and conjugated forms of the drug is not meaningful; measurement of total (free + conjugated) isoxepac rather than the separate determination of drug and metabolite might be more appropriate. So as to measure total isoxepac in urine, an efficient method of hydrolysing the glucuronide was required. During an HPLC investigation of enzymic (β-glucuronidase) and alkaline hydrolysis, a pH-dependent conversion of the 1-O-acyl glucuronide into enzyme-resistant conjugates was observed. It took place under relatively mild conditions (pH <6, room temperature), and enzyme-resistant conjugates were detected even in freshly collected urine samples. These observations cast doubt on the usefulness of β-glucuronidase for the assay of total isoxepac in urine.

In this article the rearrangement, chromatography and hydrolytic procedures for these conjugates are described, and a quantitative method for determining total isoxepac is reported.

HPLC OF ISOXEPAC AND ITS CONJUGATES: General conditions

The column was of stainless steel, 15 cm × 3 mm i.d., slurry-packed with 5 μm ODS-Spherisorb (Phase Separations Ltd., U.K.). The effluent was monitored at 280 nm using an LC 3 UV detector (Pye Unicam Ltd.). The M45 pump (Waters Associates) was run at 1 ml/min. The mobile phase consisted of 27% (v/v) acetonitrile in distilled water, adjusted to pH 2.5 with orthophosphoric acid. It was degassed before use by bubbling helium through it. Samples were injected via a Rheodyne 7120 loop injector. Peak areas are in arbitrary units.

Urine pH was adjusted to 6, 7, 8 or 9 with dilute acid or alkali. The pH was determined using narrow range pH papers (Pehanon, Camlab Ltd.). Samples were incubated at 37° and aliquots taken for analysis at regular intervals. For enzyme studies the pH was 5.2, with β-glucuronidase/aryl sulphatase Type H2 (*Helix pomatia*; Sigma Chemical Co.), at a final concentration relative to β-glucuronidase of 0.07 E.U./ml. The aldonolactone inhibitor, prepared as described by Levy [2], was present at a final nominal concentration of 20 mM.

ASSAY OF URINE FOR TOTAL ISOXEPAC: General conditions

Samples were obtained from healthy volunteers dosed with 800 mg isoxepac orally. Urine samples were stored frozen at -20° until analyzed. Isoxepac is unstable in the presence of strong sunlight, and samples were protected by covering them with aluminium foil.

To 1 ml of sample, 1.5 ml of 0.1 M NaOH and 100 μl of i.s. (a chloro analogue) were added. Samples were then incubated at 20° for 30 min, and analyzed on a fully automatic system. The HPLC pump and UV detector were as above; the latter had a Schoeffel FS 970 fluorimeter (excitation, 270 nm; emission, >398 nm) connected in series. A WISP autoinjector (Waters Asociates) was used. The output from the detectors was processed by two SP 41200 Computing Integrators (Spectra Physics Ltd.), furnishing peak areas. The chromatographic conditions were as above, except that the column was 10 cm × 2 mm and the ODS-Spherisorb was 10 μm.

RESULTS AND DISCUSSION

Chromatography, rearrangement and hydrolysis of isoxepac conjugates

In the absence of an authentic sample of the 1-*O*-acyl glucuronide, a chromatographic separation of the drug and its conjugate(s) was developed on a sample of human urine known to contain both drug and metabolite(s). Both contain a carboxyl group; hence a pH value

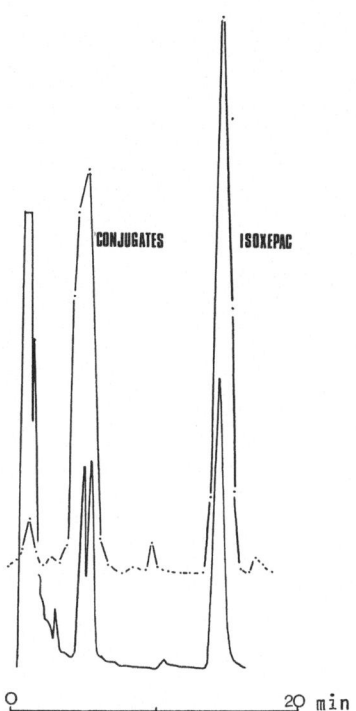

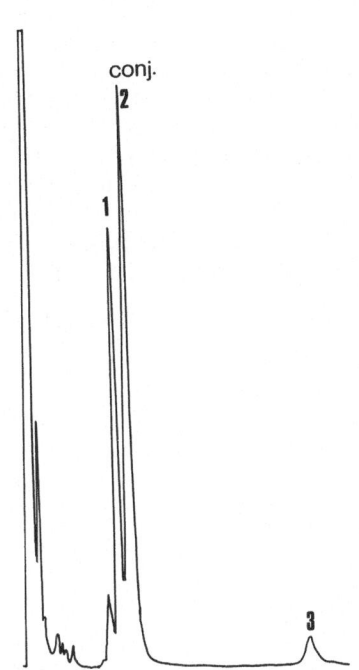

Fig. 1. HPLC of a sample of
radiolabelled isoxepac and
its glucuronides obtained
from human volunteer urine,
showing UV and radioactivity
profiles (30 sec fractions).
·–·–·, radioactivity;
———— , UV.

Fig. 2. HPLC of a freshly collected
sample of human urine containing
isoxepac (3) and its glucuronides
– 1 & 2; 2 = the presumed 1–O–acyl
glucuronide, R_t 6 min. On standing,
the latter decreases and peaks at R_t
5.6, 5.1, 4.8 & 4.7 appear or increase. Total analysis time 20 min.
(R_t = t_R = retention time.–Ed.)

of 2.5 was chosen for the mobile phase, in order to suppress ionization and get retention. A urine sample was obtained from a healthy
male volunteer, dosed orally with radiolabelled isoxepac (200 mg,
50 µCi). Using this sample conjugates were identified by monitoring
fractions from the column for radioactivity. Peaks co-chromatograph-
ing with the ^{14}C label were clearly visible on the UV trace (Fig. 1).
As would be expected for ester glucuronides, hydrolysis with alkali
resulted in the disappearance of conjugate(s), and a concomitant
increase in levels of the free drug. The presence of more than one
peak in the 'conjugate' portion of the chromatogram appeared incon-
sistent with the findings of our previous work showing the 1–O–acyl
glucuronide as the major human metabolite of isoxepac [1]. So the
possibility of artefact formation was investigated (Fig. 2), thus.-

Chromatography of various fresh samples derived from volunteers

demonstrated the presence of a variable number of drug-related peaks
in the conjugate portion of the chromatogram (Fig. 2), all of which on
alkaline hydrolysis yielded free drug.   The number and proportions
of these peaks were seen to change over a period of several hours in
samples stored at room temperature. In urines incubated at pH values
between 6 and 9 at 37° two processes were observed.   Firstly there
was a pH-dependent hydrolysis of all the conjugates to produce free
isoxepac, the rate being faster at high pH (Fig. 3).   Secondly there
was conversion of the compound giving the peak of R_t 6 min into a
number of other compounds, the rates of formation of which also increased with increasing pH.   After 3 h at pH 7 (37°) and 1 h at pH 8

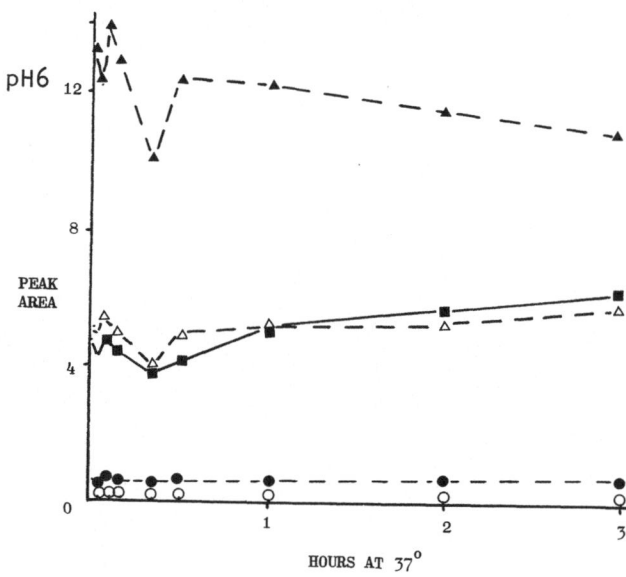

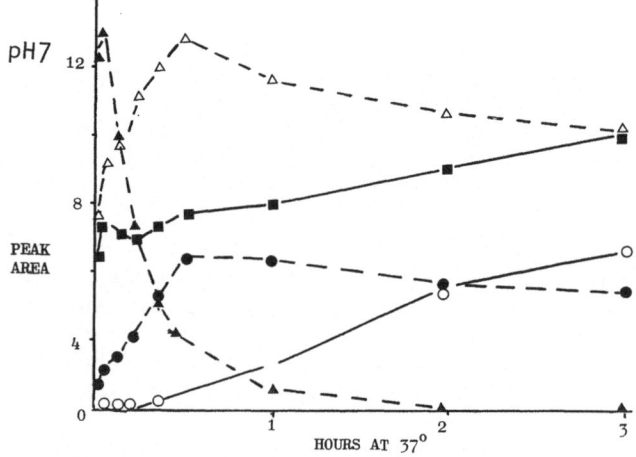

Fig. 3. Rates of hyd-
rolysis and rearrange-
ment at pH 6 and 7. At
pH 6 both hydrolysis
and rearrangement are
slow; at pH 7 both pro-
cesses are faster. The
symbols signify peaks
of retention: ▲, 6 min;
●, 5.1 min; o, 4.6 and
4.8 min; ■, isoxepac,
20 min.

(37°) the relative proportions of all the conjugate peaks remained constant, and the only observed change was hydrolysis of all the conjugates to isoxepac. At pH 9 (37°) hydrolysis was so rapid that the rearrangements observed at lower pH could not be followed by our methods.

The effects of β-glucuronidase on the rearranged products were also investigated. Both 'fresh' urine samples and samples pre-incubated at pH 8 (37°) for 3 h were treated with β-glucuronidase. The enzyme hydrolysed only the unrearranged conjugate (R_t 6 min), and was without effect on the rearranged products. Non-enzymic hydrolysis in a control incubation was negligible for all the conjugates. In a further experiment the enzyme was inhibited using the aldonolactone preparation of Levy [2]. This gave 85% inhibition of the hydrolysis of the unrearranged conjugate. The peak at 6 min is therefore the 1-O-acyl glucuronide of isoxepac.

Both rearrangement and hydrolysis to free drug occur under relatively mild conditions, such as prevail in the bladder. Analysis of samples immediately after collection gave variable quantities of of the 1-O-acyl glucuronide and enzyme-resistant conjugates, with low levels of free drug. The small quantities of free drug found in freshly collected samples suggest that the measurement of free drug and conjugates separately may be of limited value (see also [3]). The use of β-glucuronidase to obtain total isoxepac (drug + metabolite) will provide inaccurate and variable results, even if further rearrangement is prevented by acidifying the sample on collection. Because of this we have developed an assay for total isoxepac based on alkaline hydrolysis.

Quantitative determination of total isoxepac in urine

A number of chromatographic assays have been described for the determination of isoxepac, using GC-FID [4], HPLC-UV [5] and HPLC-fluorescence [6]. All of these use a pre-chromatographic extraction step to concentrate the drug. The relatively high levels of isoxepac present in urine (1 μg - 2 mg/ml) allow samples to be analyzed without recourse to extraction procedures. Thus sample preparation is limited to dilution of the sample with alkali (to hydrolyse conjugates) and the addition of an i.s. if required. The simplicity of the method means that an i.s. is not strictly necessary and may be omitted without loss of either accuracy or precision and with reduced analysis time. Replicate analysis (6 replicates) of samples containing isoxepac over the range 0 - 1000 μg/ml, assayed using a single point calibration standard (3 replicates at 100 μg/ml), demonstrated that the results obtained with and without i.s. are very similar (Table 1). Except at low levels (<1 μg/ml) the method is accurate and precise, with a detection limit of 0.9 μg/ml for UV and 0.2 μg/ml for fluorescence. The replicate analysis of conjugate-containing samples (6 replicates) over the range 0 - 1000 μg/ml shows the hydrolysis to be reproducible. Fig. 4 shows a typical chromatogram.

Table 1. Assay method for total isoxepac in urine: accuracy and precision [standard deviation (S.D.) & coefficient of variation (C.V.), with and without internal standard (i.s.)]. Note that the blank has not been deducted. Each result is the mean of 6 determinations for the same sample on separate occasions. Rec. denotes mean recovery.

Isoxepac spike, µg/ml	Isoxepac found, with an i.s.			Isoxepac found, no i.s.		
	Mean ± S.D.	Rec., %	C.V., %	Mean ± S.D.	Rec., %	C.V., %
0	0.5 ±0.3	–	–	0.5 ±0.3	–	–
1.0	1.3 ±0.2	130	15.4	1.3 ±0.2	130	15.4
3.0	3.5 ±0.4	116	11.4	3.6 ±0.4	120	11.1
10	10.3 ±0.5	103	4.9	10.6 ±0.6	106	5.7
30	33.9 ±1.8	113	5.3	34.9 ±1.5	116	4.3
100	104.1 ±2.9	104	2.8	107.2 ±2.6	107	2.4
300	316.1 ±9.4	105	3.0	318.9 ±8.1	106	2.5
1000	1037 ±44	103	4.2	1051 ±29.2	105	2.8

Both UV and fluorescence detection may be used; however in this application we find UV detection to be most useful. The higher sensitivity of fluorescence is not required, and at high levels (>300 µg/ml) the fluorescence yield is non-linear.

CONCLUDING COMMENTS

The hydrolysis of isoxepac glucuronides to isoxepac casts doubt on the usefulness of determining free and conjugated forms separately. Indeed, some if not all of the free drug encountered in urine samples may result from the hydrolysis of glucuronides. This may be a general phenomenon for ester glucuronides, as similar results have been reported for ketoprofen, naproxen and probenicid [3]. The formation of enzyme-resistant compounds in samples prior to collection demands the use of alkaline hydrolysis if total drug levels are to be determined.

The pH-dependent transformations observed for isoxepac glucuronides may be similar to the pH-dependent transacylation of the 1-O-acyl glucuronide of bilirubin, to give 2-, 3- and 4-O-acyl forms [7]. In the case of probenecid, 1-, 2-, 3- and 4-O-acyl glucuronides have been isolated [8]. Both transacylation [9] and the formation of α- and β-furanoside and α-pyranoside forms of the ester glucuronide of clofibric acid [10] have been proposed. [Cf. J. Caldwell et al. in the preceding article, #C-2.-Ed.]

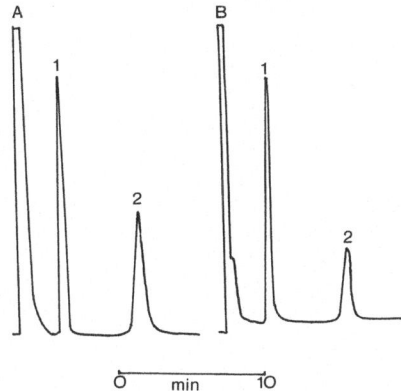

Fig. 4. HPLC traces, with fluorescence (**A**) or UV (**B**) detection, showing peaks for (1) isoxepac, 100 µg/ml, and (2) i.s., 100 µg/ml, with the assay conditions developed for total isoxepac (drug + metabolite) in human urine.

   Such transformations may be a general phenomenon of ester gluc-uronides, and appropriate precautions should be taken in their assay. The ability of RP-HPLC to separate such closely related isomeric glucuronides is noteworthy, and demonstrates the usefulness of this technique in such studies.

   We have shown that with alkaline hydrolysis, which is rapid and efficient, a precise, accurate and specific analytical method for total isoxepac in urine is feasible.

References

1. Illing, H.P.A. & Fromson, J.M. (1978) *Drug. Metab. Dispos. 6,* 510–517.
2. Levy, G.A. (1952) *Biochem. J. 52,* 464–472.
3. Upton, R.A., Buskin, J.N., Williams, R.Z., Holford, N.H.G. & Riegelman, S. (1980) *J. Pharm. Sci.* 69, 1254–1257.
4. Bryce, T.A. & Burrows, J.L. (1978) *J. Chromatog. 145,* 393–400.
5. Slack, J.A. (1980) *J. Chromatog. 221,* 431–434.
6. Hundt, H.K.L. & Brown, L.W. (1981) *J. Chromatog. 225,* 482–487.
7. Blanckaert, N. Compernolle, F., Leroy, P., van Houtte, R., Fevery, J. & Heirwegh, K.P.H. (1978) *Biochem. J. 171,* 203–214.
8. Eggers, N.J. & Doust, K. (1981) *J. Pharm. Pharmacol. 33,* 123–124.
9. Sinclair, K.A. & Caldwell,J. (1981) *Biochem. Soc. Trans. 9,* 215.
10. Hignite, C.E., Ischanz, C., Lemons, S., Wiese, H., Azarnoff, D.L. & Hoffman, D.H. (1981) *Life Sci. 28,* 2077–2081.

also (added by Senior Editor):

11. Illing, H.P.A. & Wilson, I.D. (1981) *Biochem. Pharmacol. 30,* 3381–3384.

#NC(C)

NOTES and COMMENTS relating to

Investigation of conjugates

Comments relating to particular contributions:

#C-1 & #NC(C)-1, p. 197

#NC(C)-1

Analytical case history

A Note on

ISOLATION AND IDENTIFICATION OF A DRUG CONJUGATE

M.P. Harrison

Safety of Medicines Department
ICI Pharmaceuticals Division
Alderley Park, Macclesfield SK10 4TG, U.K.

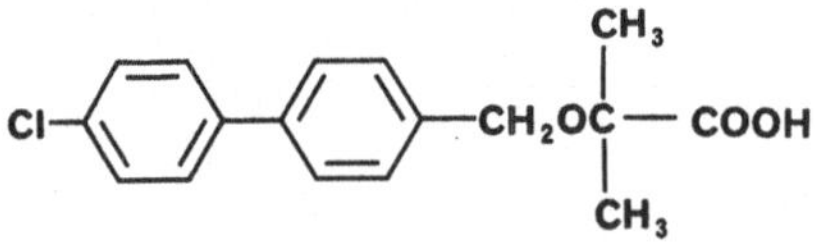

ICI 55,897 ('Clozic') was developed by ICI for potential use as a treatment for rheumatoid arthritis. Metabolism studies in animals and man using ^{14}C-labelled Clozic showed marked species variation in the route and rates of excretion after oral administration (Table 1). However, in all species the compound is excreted as a single conjugate of unchanged Clozic. Early attempts to characterize this conjugate using ^{14}C-labelled Clozic were hampered by the fact that the conjugate hydrolyzed spontaneously and rapidly at pH values more alkaline than 3. This made conventional enzymic tests for conjugates difficult to interpret, as control samples also showed rapid hydrolysis. In fact, urine samples have to be acidified immediately after collection to keep the total conjugate intact. The Clozic conjugate was eventually isolated from hamster urine by solvent extraction at acid pH and purified by RP-HPLC. Chemical identification as a glucuronide has been made on the basis of both proton and ^{13}C NMR analysis.

Pooled hamster urine samples were acidified immediately on collection with M HCl and the total Clozic-related material in the sample (250 ml) was determined by measurement of the ^{14}C content. This was equivalent to ~60 mg of ^{14}C-labelled Clozic, and RP-HPLC analysis (Fig. 1) showed that most of the ^{14}C material chromatographed as a single peak which eluted before Clozic. The total urine sample was extracted 3 times with 2 vol. of ethyl acetate, which extracted 80-90% of the total ^{14}C material. The combined extracts

Table 1. Excretion of ^{14}C after oral administration of ^{14}C-labelled ICI 55,897 to various species. Urinary and faecal values are % of dose.

Species	Urine	Faeces	$t_{\frac{1}{2}}$ Elimination
Mouse	10%	90%	20 h
Rat	5%	95%	4-6 days
Hamster	80%	20%	12-36 h
Marmoset	80%	15%	8 h
Cynomolgous Monkey	80%	15%	12 h
Man	85%	–	3-6 days

Fig. 1. RP-HPLC runs on hamster urine containing the ^{14}C metabolite and also, in **2** *(left)*, added parent compound (^{14}C-Clozic).

Column: 20 cm × 4.5 mm i.d. packed with 5 μm Spherisorb ODS.
Eluent: methanol/ 0.5% trifluoroacetic acid (TFA), 4:1 by vol., 1.5 ml/min.
Detection: 285 nm, and cpm by a Berthold LB 503 LC radioflow monitor.
Injection: 250 μl.

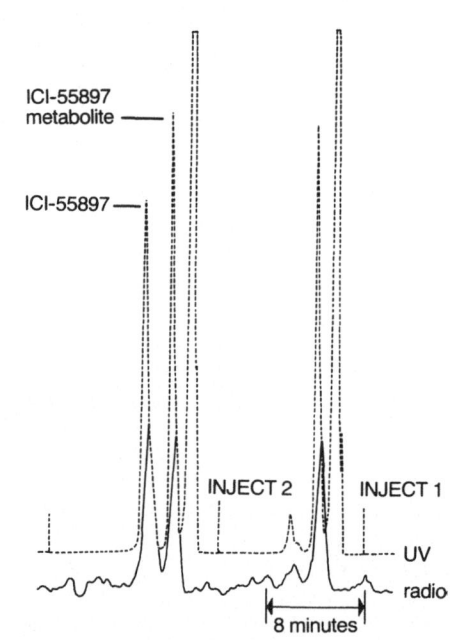

were dried down, and the residue dissolved in 3 ml of methanol/0.2% TFA (1:1 by vol.). The sample was then subjected to preparative HPLC, firstly with 3 separate runs using a 25 cm × 9 mm i.d. column packed with 60-75 μm Merck LiChroprep RP18, with isocratic elution as in the Fig. 1 legend. The fractions detected as radioactive peaks were collected and the total combined. The HPLC solvent was removed by freeze-drying, and the ^{14}C material recovered was rechromatographed using gradient elution (Fig. 2). The product after drying down was again dissolved in a small volume of 1:1 methanol/ TFA and loaded onto the same column isocratically with the same mixture. Gradient elution as in the legend to Fig. 2 was again performed, and the material detected as a ^{14}C peak was collected and recovered by freeze-drying. The sample obtained (~25 mg) was then analyzed by both proton and ^{13}C NMR.

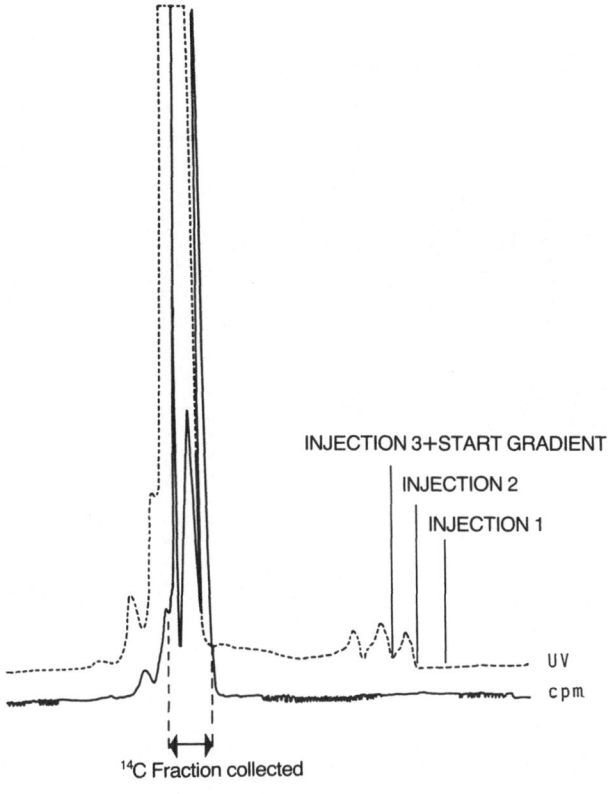

Fig. 2. Final RP–HPLC purification of the ^{14}C metabolite.

Column: 250 × 9 mm i.d. 60–75 µm LiChroprep RP18.
Eluent: linear gradient, 50 → 100% methanol with 0.1% TFA present.
Flow-rate 8 ml/min.
Injections (methanol/ 0.2% TFA, 1:1): 1 ml of sample, then (2,3) 1 ml twice more, with gradient start after 3.
Detection as in Fig. 1 legend; ^{14}C with auto-ranging, 3×10^3–3×10^4 cps.

INJECTION 3+START GRADIENT

INJECTION 2

INJECTION 1

UV

cpm

^{14}C Fraction collected

        Proton spectra were obtained at 90 MHz (and ^{13}C spectra at 22.5 MHz) using a Jeol FX 90Q Fourier Transform NMR spectrometer operating in the pulse mode with tetramethylsilane as i.s. In the examples shown the samples were dissolved in deutero-methanol. The proton spectrum (Fig. 3) indicates peaks as singlets at 7.3δ for the 8 aromatic protons, 4.47δ for the methylene carbon –CH_2CO and 1.55δ arising from the side-chain methyl groups. The 4-proton complex at 3.4δ can be attributed to the protons on C_2, C_3, C_4 and C_5 of a glucuronic acid ring, but the most useful information shown on the proton spectrum is probably the characteristic doublet at 5.6δ due to the anomeric proton on C_1 of glucuronic acid.

        In the absence of other information the proton NMR spectrum does not give a definitive structure for the metabolite. Rather more information is available from the ^{13}C spectra (Figs. 4 & 5). The ^{13}C spectra were obtained with proton noise decoupling and also with single-frequency off-resonance decoupling (shown as S.F.O.R.D. in Fig. 5). The carbonyl carbons at C_6 of glucuronic acid and Clozic are evident at 172.7 and 169.9 ppm. The IPSO* carbons of the aroma-

* signifies ring carbons attached to other C groups.– *Ed.*

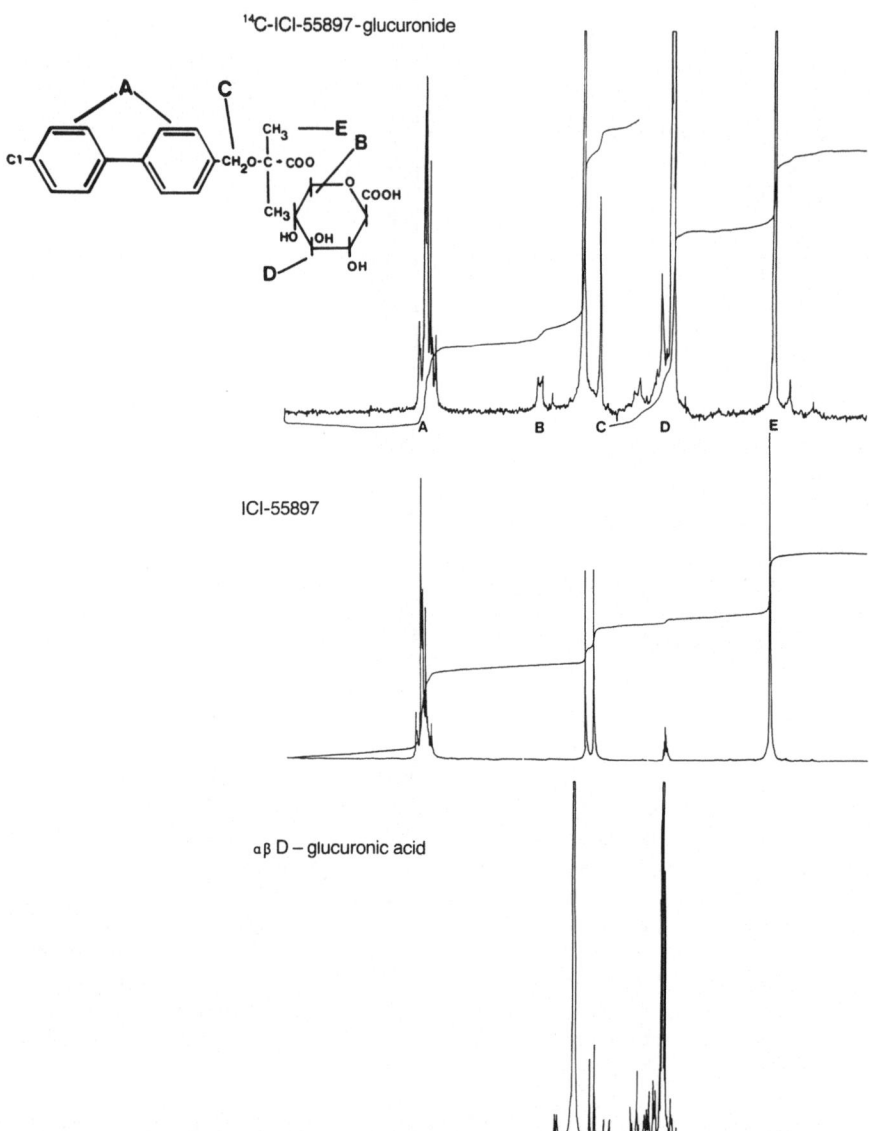

Fig. 3.   90 MHz proton NMR spectra to elucidate the structure of the conjugate of Clozic (ICI 55, 897).

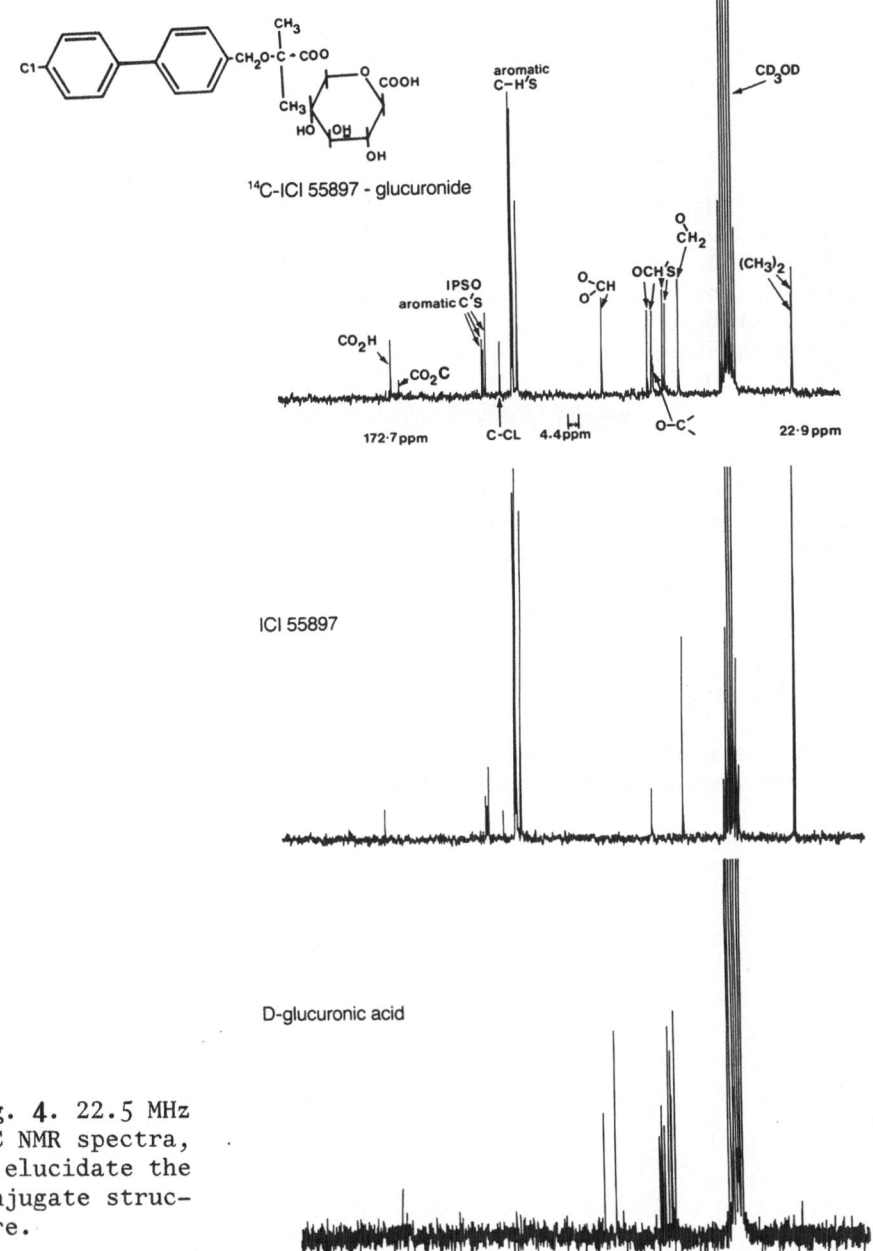

¹⁴C-ICI 55897 - glucuronide

ICI 55897

D-glucuronic acid

Fig. 4. 22.5 MHz
¹³C NMR spectra,
to elucidate the
conjugate struc-
ture.

tic rings appear at 132.3 ppm, and those bearing protons at 125–127. These latter become doublets on S.F.O.R.D.   The side-chain quater-nary carbon of Clozic is seen at 75.4 ppm, the methylene carbon at 65.9 ppm, and the two methyl groups at 22.9 and 23.2 ppm.   The

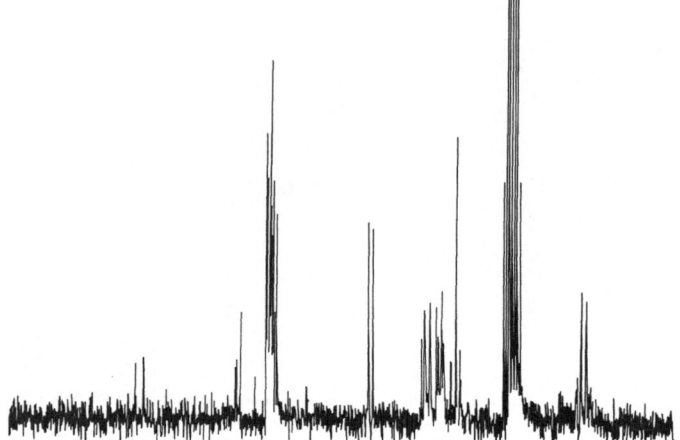

Fig. 5. 22.5 MHz
^{13}C NMR spectrum
with S.F.O.R.D.
to identify the
carbons bearing
single protons in
the Clozic con-
jugate.

glucuronide ring carbons are at 94.2 ppm (C_1) and 77.3, 75.7, 71.7 and
70.9 ppm (C_2 to C_5 respectively; all of these become doublets on
S.F.O.R.D.). These data taken together with corresponding spectra
for authentic Clozic and β-D-glucuronic acid show that the urinary
metabolite of Clozic is an ester glucuronide.

Acknowledgement

I am grateful to Mr. R. Pickford (Physical Chemistry Section)
for his assistance in obtaining and interpreting the NMR spectra.

Comments on material in #C

Comments on #C-1, J. Tomašić – CONJUGATE ISOLATION & IDENTIFICATION

Remarks by H. von Hodenberg.– I have normally tried to identify conjugates as whole molecules, e.g. by MS after HPLC isolation
and radioactivity detection, the biological samples being injected
directly onto the column. In the case of glucuronides we are mostly
successful, but with sulphates we often have problems in derivatizing
to form methyl esters. I wonder about the feasibility of identifying
sulphates as such. *Responses.*– (J. Tomašić) Paper electrophoresis ?
(L.E. Martin) In the high-voltage mode, paper electrophoresis can
indeed be of great value for identifying sulphates and glucuronides
– as has been well reviewed [1].

Concerning J. Tomašić's doubts about the feasibility of separating
diastereoisomers under ordinary TLC conditions.–*Comment by* W. Dieterle:
TLC on silica gel can indeed separate diastereoisomeric compounds effectively, e.g. warfarin-alcohols, and acetocoumarol-alcohols, besides
oxaprotiline glucuronides. (W. Ritter) We have recently separated
diastereomeric Phase I metabolites by HPTLC – and even quantitated
them by HPTLC densitometry. So a change of mode from 'simple' to 'HP'
might well be successful. *Note by Editor* (E.R.).– The ion-pair mode
of HPLC has been shown to effect separation of enantiomeric amines,
e.g. alprenolol enantiomers, with (+)-10-camphorsulphonate employed as
the pairing ion [2]. This successful outcome, which hinged on there
being a "good fit between the binding groups", was obtained under NP
conditions, with 10 μm LiChrosorb-DIOL which, in a 2-day pre-equilib-
ration with re-circulation, acquired a coating of 1-propanol that was
present in the dichloromethane mobile phase.

1. Conway, W.D. (1977) in *Drug Fate and Metabolism*, Vol. 1 (Garrett,
   E.R. & Hirtz, J.L., eds.), Dekker, New York, pp. 65-134.
2. Pettersson, C. & Schill, G. (1981) *J. Chromatog. 204*, 179-183.

Comments on #NC(C)-1, M.P. Harrison – STUDY OF A CONJUGATE

Remarks by C.R. Jones.– In view of the high water-solubility
of the compound, an RP cartridge could be a useful means of extracting it. *Response:* we have not tried this, but there is no difficulty in extracting at acid pH into ethyl acetate.

Response to B.F.H. Drenth, in the context of metabolite instability.- The metabolite is relatively stable in methanol as used to
obtain the decoupled ^{13}C-NMR spectra, which takes about 30 min. *Response to* G.R. Bourne.- In L.E. Martin's view, field-desorption (FD)
MS is useful in conjugate identification, being a valuable technique for obtaining a molecular ion; only µg of material are needed,
compared with mg for NMR analysis.

Editor's reminder about guide to conjugate types

The Tables at the end of #E below may help the reader gain perspective on types of conjugate and, in the case of glucuronides, how the
different types respond to hydrolytic treatment. Glucuronides (once
termed 'glucosuronidates', now allowed rather than insisted upon) are
normally *O*-linked and are of pyranose type, by definition.

Comments on the usefulness of conjugate isolation/assay

Responses by J. Chamberlain, C.R. Jones, M.P. Harrison.- With an
unstable conjugate as found with isoxepac (#C-3), measurement of the
intact conjugate rather than the aglycone could be misleading. There
is always a need to study the aglycone, since it occurs free in the
body at some stage and may have more than one conjugate form. One
must keep in mind the recycling of glucuronides (resorption in the
large intestine via microfloral metabolism; release of the aglycone
as an 'active' agent) and species differences. *Ref. added by Ed.*-
A pointer to the biological interest of ascertaining conjugate profiles is a change observed (for Phase I profile also) in respect of
Vitamin K_1 when warfarin was administered [1].

1. McBurney, A., Shearer, M.J. & Barkman, P. (1978) *Biochem.
   Pharmacol. 27*, 273-278.

Section #D

DETERMINATION OF PARTICULAR DRUGS AND METABOLITES

#D-1

Analytical case history

HPLC ANALYSIS OF THE MAJOR URINARY METABOLITE OF FLURAZEPAM

J.A.F. de Silva, M.A. Brooks
M.R. Hackman and R.E. Weinfeld

Department of Pharmacokinetics and Biopharmaceutics
Hoffmann-La Roche Inc.
Nutley, NJ 07110, U.S.A.

Requirement /metabolism	*Determination of urinary N_1-hydroxyethylflurazepam ('II'; excreted as glucuronide) in pharmacokinetic studies, as an alternative to blood determinations. Certain other metabolites also occur (Table 1).*
End-step	*HPLC, using C-18 silica, with 254 nm detection; aq. methanol eluent, pH 3.5. 'DPA' may be informative.*
Sample preparation	*After enzymic hydrolysis, adjust to pH 9.0, add i.s., and extract with diethyl ether; dry down and take up residue in eluent.*
Comments	*Detection limit ~100 ng/ml; poorer (~ 500 ng/ml) if normal-phase HPLC although retention time is advantageously shorter. The glucuronide runs early.*

Pharmacokinetic parameters of a drug are commonly evaluated by the determination of unchanged drug in blood or plasma. An acceptable alternative [1], especially if a non-invasive method is desired, is to use urinary excretion data for either the unchanged drug or a major metabolite. If the latter represents a significant percentage of the administered dose, its quantitation places fewer constraints on the sensitivity required of the analytical procedure, and on the sample preparation and 'clean-up' required prior to analysis. HPLC, either normal-phase (NP) or reverse-phase (RP), is ideally suited to this type of determination [2]. RP conditions have now been developed.

Flurazepam dihydrochloride, [I].2HCl, a hypnotic of the 1,4-

Table 1. Chemical structures and relative retention times of flurazepam and its metabolites determined by normal-phase (NP) or reverse-phase (RP) HPLC analysis. For the glucuronide of **II** (not tabulated) the RP relative-R_t value is 0.45.

Also tabulated (last 2 entries): flurazepam analogues used as internal standards. [$R_t \equiv$ other authors' t_R-*Ed.*]

	R group	Compound	Relative R_t NP	Relative R_t RP
I	$-(CH_2)_2-N-(C_2H_5)_2$	**Flurazepam**	0.88	0.96
I-A	$-(CH_2)_2-NH-C_2H_5$	Mono-desethyl-flurazepam	0.55	0.75
I-B	$-(CH_2)_2-NH_2$	Di-desethyl-flurazepam	2.43	0.65
II	$-CH_2-CH_2OH$	*N*-1-hydroxyethyl-flurazepam	(1.00)	(1.00)
III	$-H$	*N*-1-desalkyl-flurazepam	0.89	1.14
IV	$-H$, and 3CHOH	*N*-1-desalkyl-3-hydroxy-flurazepam	3.75	0.81
V	$-(CH_2)_2-N-(CH_3)_2$, and Cl at 2'-used as NP-HPLC i.s.		1.56	–
	$-CH_3$	Diazepam - used as RP-HPLC i.s.	–	1.88

benzodiazepine class, undergoes extensive biotransformation in man [3] by successive dealkylation and oxidation to form several metabolites (Table 1). The major urinary metabolite, **II**-glucuronide, accounts for 30-55% of an orally administered dose in a 72 h excretion period (>25% in 24 h) [4]. Methods for the quantitation of **I** and its metabolites in blood or plasma have included luminescence [4], GC-ECD [e.g. 5, 6], differential pulse polarography [5], spectrofluorodensitometry [7], RIA (8) and GC-MS [9, 10]. The present study concerns HPLC approaches for **II**, especially in the RP mode.

SAMPLE PREPARATION AND HPLC APPROACHES

For conjugate hydrolysis, 2 ml of 1 M pH 5.45 phosphate buffer and 10 µl of a glucuronidase/sulphatase preparation ('Glusulase') were added to the 1 ml urine sample, and the mixture was incubated overnight at 37°. Then, except where direct analysis was to be performed, the above-mentioned extraction [11] was performed, and the residue taken up in 1.0-4.0 ml of the mobile phase, the i.s. being added then (in 10 µl) or, with the RP mode, prior to extraction. The pH 9 extraction gives a quantitative recovery of **II**, relatively free of endogenous impurities [4].

The HPLC conditions are given in the Fig. legends for the two modes, viz. NP [11] or RP, with or without extraction. Waters Associates equipment was used (pump, 6000A; fixed-wavelength UV

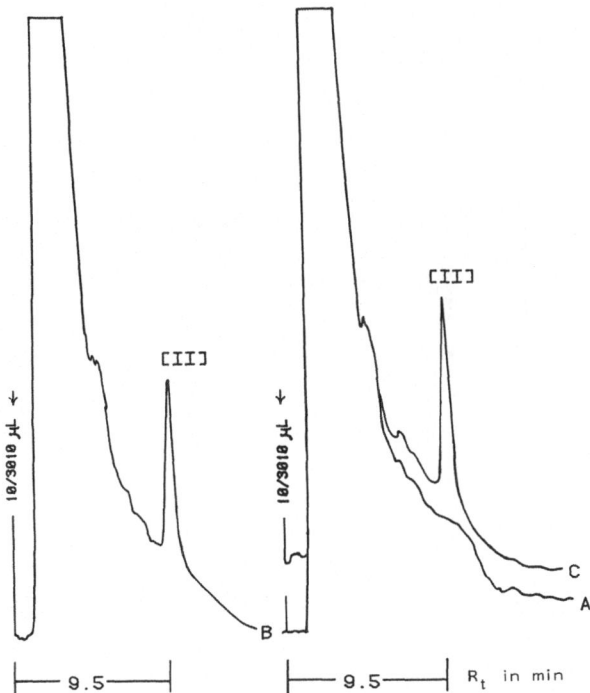

Fig. 1. Direct analysis of urine incubate by RP-HPLC. **A** (on right): control urine. **B** (on left): control urine spiked with 10 µg/ml of the metabolite **II** (Table 1). **C** (on right): patient's urine, 1-2 h post-dose, manifesting **II** at a level of 11.8 µg/ml. Sensitivity reckoned to be 1.0 µg/ml.
Column: 300 x 3.9 (i.d.) mm, stainless steel, containing 10 µm Bonda-pak C-18 (Waters), pre-packed; directly before it was a 20 x 3.9 mm guard column, stainless steel, containing 15-40 µm Bondapak C-18, hand-packed. Mobile phase: methanol-5 mM pH 3.5 phosphate buffer, 55:45 by vol.; per 1 of mobile phase, 2.5 ml of 1 M buffer (KH_2PO_4/ H_3PO_4) was diluted to 450 ml with dist. deionized water, and methanol added to 1 l. Flow rate 1.5 ml/min (2300 psi). Detection at 254 nm. Injection: 10 µl, at point denoted (in all Figs.) ↓. Diazepam as i.s.

dual channel detector with 254 nm filter, 440; automatic sample injec-tor, WISP 710B). Mobile phases were vacuum-degassed with ultrasonica-tion for ~5 min before use. DPA was performed with a Model 174 polaro-graphic analyzer which controlled an electrochemical detector, Model 310 (EG & G-PARC); see also Fig. 4 legend.

Comparison of different assay conditions

Direct analysis of the incubate (Fig. 1) obviates sample manipu-lation, but with the penalty of low sensitivity and poor chromato-

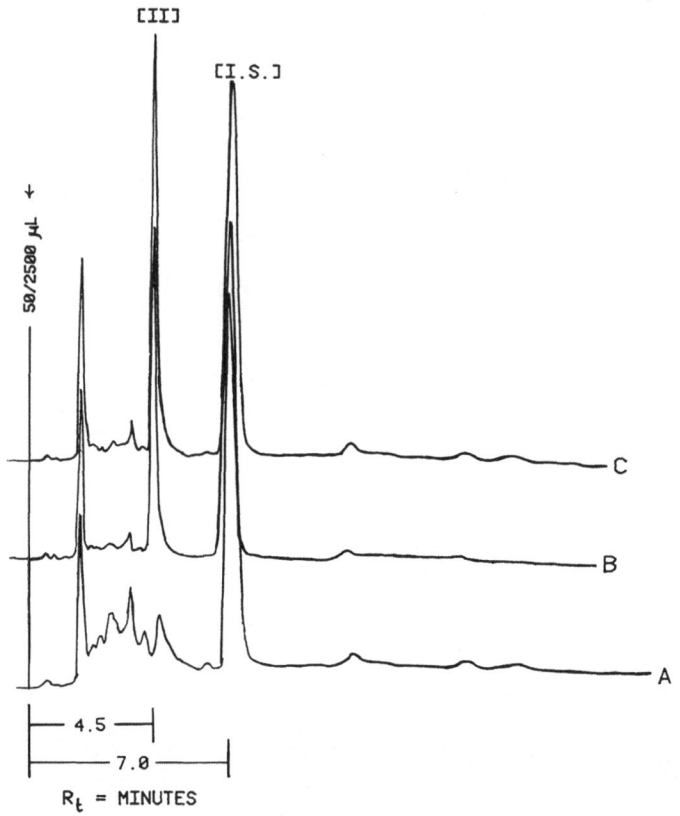

Fig. 2. Analysis of extract from urine incubate by NP-HPLC.
A: control urine, as in Fig. **1**, A. **B**, spiked urine, as in Fig.1, B.
C, patient's urine, as in Fig. 1, C. For the metabolite **II** (Table 1)
the sensitivity is reckoned to be 0.5-1.0 μg/ml.
Column: 250 × 4.6 (i.d.) mm, containing 10 μm silica gel (Partisil PXS,
Whatman Inc.). Mobile phase: a 500:25 (by vol.) mixture of dichloro-
methane and of methanol-water-conc. NH_3 (150:9:1 by vol.). Flow rate
1.5 ml/min (650 psi). Detection at 254 nm.
Injection: 50 μl, from 2.5 ml of extract. **V** (Table 1) served as i.s.

graphy due to endogenous impurities. The cleaner sample obtainable
by ether extraction, when analyzed by NP-HPLC [11], still showed
small amounts of impurities (Fig. 2) which depressed the sensitivity
compared with RP-HPLC (Fig. 3), although the run time was only 10 min
compared with 20 min. The respective retention times were 4.5 and
9.5 min for **II**. Both systems were efficient as judged by peak shape.

FINDINGS WITH THE RP-HPLC ASSAY AS ADOPTED

Peak-height ratio calibration curves were determined for extracts corresponding to 0.10-75.0 μg of II/ml of urine. Linear least-

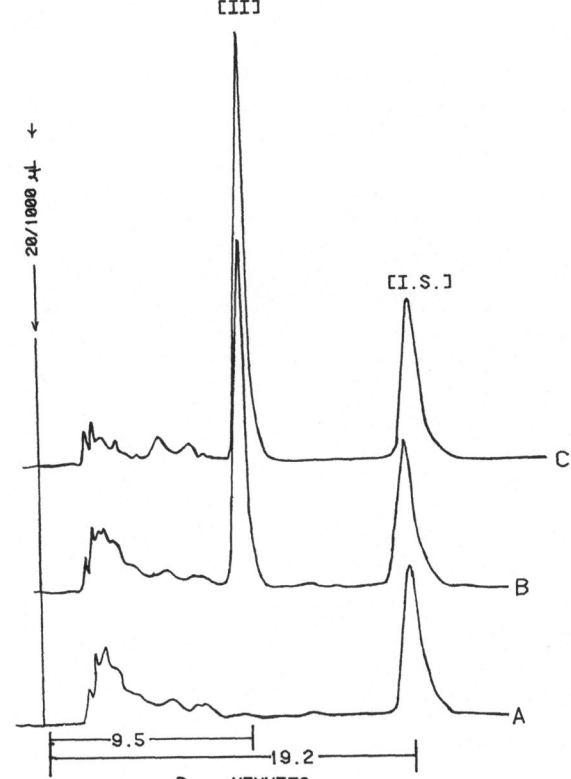

Fig. 3. Analysis of urine–incubate extract by RP–HPLC. **A, B & C** as in Figs. 1 & 2. For II the sensitivity is reckoned to be 100 ng/ml. Column, mobile phase and other conditions as in Fig. 1. Injection: 20 µl, from 1.0 ml of extract. Diazepam served as i.s.

squares regression analysis gave   r = 0.9996 and C.V.s 8.36→1.77 (n = 3 at each of 11   concentrations) with   average C.V. = 3.4% (intra-assay precision); overall recovery >95% over the entire range.   The chromatographic resolution conferred specificity for II in the presence of intact flurazepam and minor metabolites (cf. Table 1).

When the method was compared with the NP method on urine samples taken up to 72 h after a normal volunteer had taken a single 30 mg oral dose of I.2HCl (Dalmane), there was excellent agreement in the excretion-rate values (max. at 1-2 h post-dose, 2.4 mg/h; below 0.1 mg/h beyond 12 h).   Thus either method can serve for pharmacokinetic work.

Hydrolysis of II-glucuronide followed by the direct RP method

Determination of the glucuronide *per se* was precluded by the lack of an authentic standard; so its enzymic hydrolysis was studied as a step in method development.   Specificity as achieved by chromatographic separation was confirmed (Fig. 4) by complementing the 254 nm detection with subsequent passage through an electrochemical detector (DPA; see Fig. 4 legend).   This revealed 2 extra components (probably minor metabolites; cf. Table 1) in the solvent front as monitored by UV, further testifying to the merit of DPA for specific HPLC detection [12, 13].

Fig. 4. RP-HPLC (dual detection),
showing enzymatic hydrolysis of
II-glucuronide to yield II, on
sample 1-2 h after a 30 mg dose:
30-min urine incubate (500 µl
methanol finally added to in-
activate the enzyme). Monitoring
(**A**) by differential pulse amp-
erometry (DPA) as well as (**B**)
by UV absorption (usual 254 nm).
Column etc. as in Fig. 1, but
mobile phase consisted of 7.5 mM
pH 3.5 acetate -methanol, 1:1 by
vol.; 0.9 ml/min (1500 psi).
A: -0.760 V *vs*. Ag/AgCl with
0.5 µa range setting (specific
for azomethine group [5, 12, 13]).
Injection: 50 µl, from 1.1 ml
of incubate; urine as in Fig. 1.

The II-glucuronide was ~30%
hydrolyzed; cf. 45 min, ~50%;
90 min, ~95%. Overnight incubation was adopted for convenience.

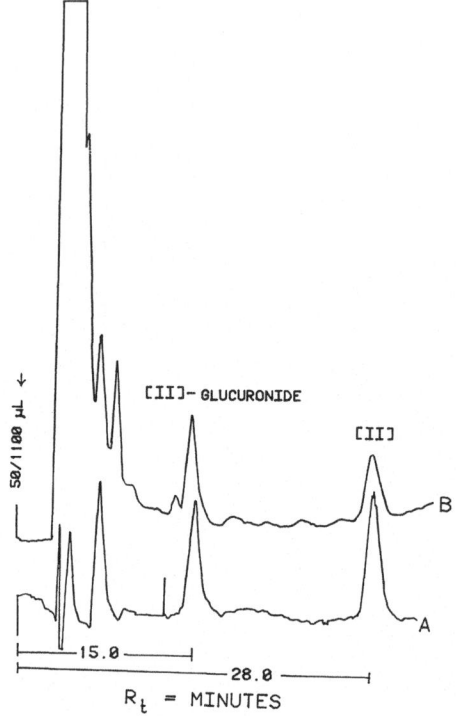

Acknowledgement

The authors thank Miss N. Bekesewycz for preparing this manuscript.

References

1. Federal Register (1977) *42* (5), 1624-1653.
2. de Silva, J.A.F. (1981) in *Trace-Organic Sample Handling* (Vol.
    11, this series; Reid, E., ed.), Horwood, Chichester, pp. 192-204.
3. Schwartz, M.A. & Postma, E. (1970) *J. Pharm. Sci. 59*, 1800-1806.
4. de Silva, J.A.F. & Strojny, N.(1971) *J. Pharm. Sci. 60*, 1303-1314.
5. de Silva, J.A.F., Puglisi, C.V., Brooks, M.A. & Hackman, M.R.(1974)
    J. Chromatog. 99, 461-483.
6. Peat, M.A. & Kopjak, L. (1979) *J. Forens. Sci. 24*, 46-54.
7. de Silva, J.A.F., Bekersky, I. & Puglisi, C.V. (1974) *J. Pharm.
    Sci. 63*, 1837-1841.
8. Glover, W., Earley, J., Delaney, M. & Dixon, R. (1980) *J. Pharm.
    Sci. 69*, 601-602.
9. Clatworthy, A.J., Jones, L.V. & Whitehouse, M.J. (1977) *Biomed.
    Mass Spectrometry 4*, 248-254.
10. Miwa, B.J., Garland, W.A. & Blumenthal, P. (1981) *Anal. Chem. 54*,
    793-797.
11. Weinfeld, R.E. & Miller, K.F. (1981) *J. Chromatog. 223*, 123-130.
12. Brooks, M.A. (1980) in *Electroanalysis in Hygiene, Environmental,
    Clinical and Pharmaceutical Chemistry* (Franklin-Smyth, W., ed.),
    Elsevier, Amsterdam, pp. 287-298.
13. Hackman, M.R. & Brooks, M.A. (1981) *J. Chromatog. 222*, 179-190.

#D-2

Analytical case history

QUANTIFICATION OF METABOLITES OF AMINOPYRINE AND ANTIPYRINE IN PLASMA AND URINE

J.B. Houston and J.C. Rhodes

Department of Pharmacy
University of Manchester
Manchester M13 9PL, U.K.

Requirement /metabolism	*Assay for metabolites of aminopyrine (AP) in the demethylation pathway, and preferably for AP too.*	*Assay for metabolites of antipyrine (AN) formed by parallel pathways, and preferably for AN too.*
End-step	*RP–HPLC, 270 nm detection; pH 7 buffer containing methanol. A suitable i.s. is* p-iodoacetanilide.	*RP–HPLC, 270 nm detection; pH 5 buffer containing some acetonitrile and methanol. AP as i.s.*
Sample preparation	*Chloroform extract dried down, and residue taken up in small vol. (5-fold concentration) for HPLC.*	*Conjugates hydrolyzed by acid or enzyme, depending on which metabolite. Then as in AP procedure.*
Comments	*The procedure is rapid and straightforward; 0.1 µg/ml urine or plasma is detectable, well below levels observed. Saliva can also be assayed.*	*Problems due to instability of liberated metabolites and to adsorption. Eluent optimization critical for resolution. Procedure very satisfactory.*

Both aminopyrine and antipyrine serve as test compounds for assessing hepatic microsomal activity *in vivo* [1, 2]. Several published assays [3-6] allow their plasma concentrations to be measured over a time period of 4-5 times their half-life. However, the usefulness of the two drugs would be much increased if routine analytical methods were available to quantify their metabolites and hence allow pharmacokinetic studies on the individual metabolic pathways rather than on the gross behaviour of the parent drug.

Aminopyrine (AP): 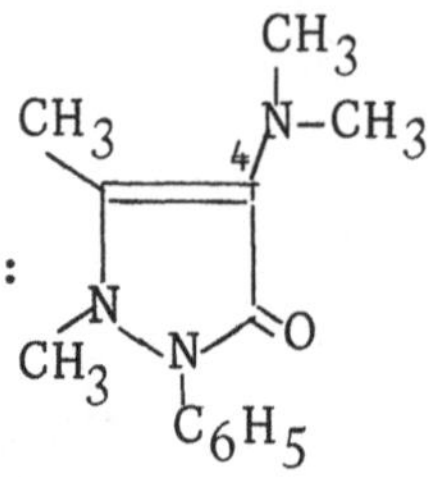   Antipyrine (AN):

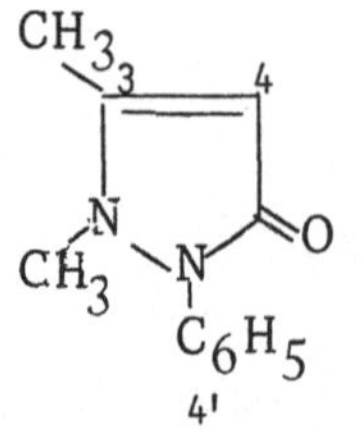

The metabolic fate of both AP and AN is well documented. AP is
is metabolized mainly at the 4-amino position, where it undergoes
two successive demethylations to give monomethylaminoantipyrine
(methylamino-AN) and aminoantipyrine (amino-AN);   the latter metabolite is *N*-acetylated (*N*-Ac-amino-AN).   AN is metabolized via four
parallel pathways to give 4-hydroxyantipyrine (4-OH-AN),   *methyl*^{3}-
hydroxyantipyrine (3-OHMe-AN), *methyl*-norantipyrine (nor-AN) and 4,
4'-hydroxyantipyrine (4'-OH-AN); also 4'-dihydroxyantipyrine (4,4'-OH-AN).

In developing routine assays for AP and AN metabolites in
urine, in the case of AP we aimed at applicability to plasma also,
as it contains metabolites of AP in appreciable concentration.   Its
AN metabolite concentrations are very low and of little pharmacokinetic consequence.   As the standard test doses of both AP and AN are
high (0.5-1 g), the urinary metabolite levels are substantial. Hence
assay sensitivity is not a problem provided that recovery from biological fluids is reasonable. RP-HPLC with a suitable column (legend
to Fig. 1) was a natural choice of approach, with specific yet rapid
separations through eluent optimization to achieve large resolution
factors and relatively low capacity factors.   AP, AN and their metabolites all have molar extinction coefficients of ~10,000 in the
range 230-270 nm; at higher wavelengths endogenous interferences are
minimized. (For sources of authentic compounds, see *Acknowledgements*).

RESOLUTION OF AP AND AN METABOLITES BY RP-HPLC

Resolution of AP metabolites was pH-dependent (Fig. 1, **A**). At
pH 7, selectivity was maximal for AP, the above-mentioned 3 metabolites, and 4-OH-AN (usually a minor metabolite of AP).   With the
organic modifier, methanol, at 25%, resolution factors were >2 (Fig.
1, **B**).   With these conditions, iodoacetanilide proved to be a good
internal standard with a capacity factor of 16, giving a total run
time of 22 min (Fig. 2).

Eluent pH is critical to the resolution of AN metabolites, the
retention of nor-AN being markedly influenced (Fig. 3, **A**). Only at
pH 5 is this metabolite adequately resolved from 3-OHMe-AN and the
normally minor metabolite 4'-OH-AN. Surprisingly, 4-OH-AN is retained longer than the parent drug; its peak shape is poor unless the
eluent contains acetate ions, hence an acetate buffer is advisable
(not used in the Fig. 3A study, because of the pH range needed).

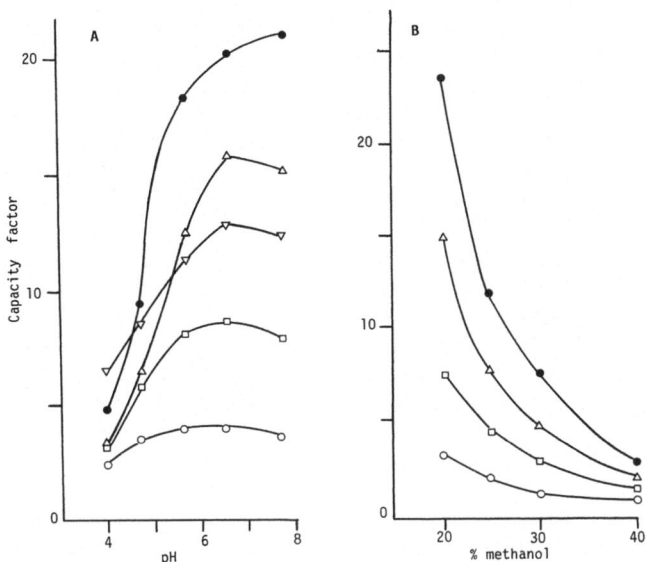

Fig. 1. HPLC capacity factors for aminopyrine and its metabolites, as influenced by: **A**, pH of eluent; **B**, methanol content of eluent. Symbols (for abbreviations, see text): AP, ●; methylamino-AN, △; amino-AN, □; N-Ac-amino-AN, ○; 4-OH-AN, ▽. Detection at 270 nm. Column: 100 × 5 mm, 'short alkyl' (Hypersil-5 SAS; Shandon, Runcorn). Eluent: **A**, methanol-0.2 M acetate buffer, 20:80 by vol., the proportion of ammonium acetate being varied to set the pH; **B**, methanol-0.05 M pH 7.0 phosphate buffer. Flow-rate 1 ml/min (as in all experiments). Sample injection: 20 µl.

[*Editorial note.*- The abbreviations for metabolites are different from those used by the authors: thus 3HM (as conventionally used) is replaced by 3-OHMe-AN. For chemical correctness, *methyl*^{3} is used as a prefix.]

Fig. 2. A chromatogram for antipyrine and its metabolites, isolated from urine as described in the text. Peaks: 1, N-Ac-amino-AN; 2, amino-AN; 3, methylamino-AN; 4, AP; 5, i.s. (see text). Column: as for Fig. 1. Eluent: methanol-0.05 M pH 7.0 phosphate buffer, 25:75 by vol. Detection at 270 nm.

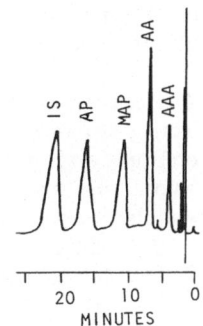

The concentration of AN in urine in the post–dose period (48 h) is only ~10% that of its individual metabolites. Adequate resolution of AN from both 4-OH-AN and nor-AN could be achieved with methanol- or acetonitrile-buffer eluents only when the organic content was low, which resulted in unacceptably high capacity factors (Fig. 3B). The inclusion of methanol as a secondary organic modifier to acetonitrile –buffer mixtures reduced the capacity factors to reasonable values without loss of resolution (resolution factors >2).

A convenient internal standard for the AN metabolite assay is AP, which has a capacity factor of 18 under the above conditions, giving a total run-time of 25 min (Fig. 4). The fourth major AN metabolite (4,4'-OH-AN) is also adequately resolved (capacity factor 6) in the final phase system; but its instability precluded its recovery from urine, as is discussed below.

WORK–UP PROCEDURE FOR AMINOPYRINE METABOLITES

There is no instability problem, and there is good extraction at pH >8. At pH 14 the recoveries relative to iodoacetanilide were AP, 102%, N-Ac-amino-AN 98%, amino-AN 70%, methylamino-AN 48%. The following procedure was adopted: to plasma (0.5 ml), saliva (0.5 ml) or urine (1 ml), NaOH (2M, 0.25 ml), iodoacetanilide and chloroform (10 ml) were added, and the mixture shaken (10 min). After centrifugation to separate the phases, the aqueous phase was aspirated and the organic layer removed with a Pasteur pipette and evaporated to dryness under nitrogen. The AP metabolites were reconstituted in methanol (25 µl) and diluted with water (75 µl) prior to taking 20 µl for HPLC under the conditions stated in the Fig. 4 legend.

WORK–UP PROCEDURE FOR ANTIPYRINE METABOLITES

Problems became manifest that did not arise with AP metabolites. As shown by incubation time studies (none hitherto reported) done on 50 µg/ml solutions in dist. water, AN and 3-OHMe-AN survive well, and likewise 4'-OH-AN; but losses occur with nor-AN (8% after 2 days at 37°), 4-OH-AN (20% in 3 days) and especially 4,4'-OH-AN (73% in one day at 20°, 20% in 3 days at 4°). This inherent instability together with the polarity of the AN metabolites made the selection of pH for solvent extraction critical. Based on data for metabolite extractability into chloroform (Fig. 5), pH 5 was finally adopted as a compromise, taking account of the problem of endogenous contaminants.

Each AN metabolite occurs in the urine mainly as conjugates, a form in which storage at −20° entails no degradation. The sulphate and glucuronide conjugates must be hydrolyzed prior to extraction and HPLC, avoiding aglycone losses due to concomitant degradation. Fig. 4 shows chromatograms for the same urine with the two approaches tried, viz. 2 M HCl for 30 min at 100°, and incubation at 37° for 180 min in 0.1 M pH 5 acetate buffer containing β-glucuronidase (Sigma Type H1).

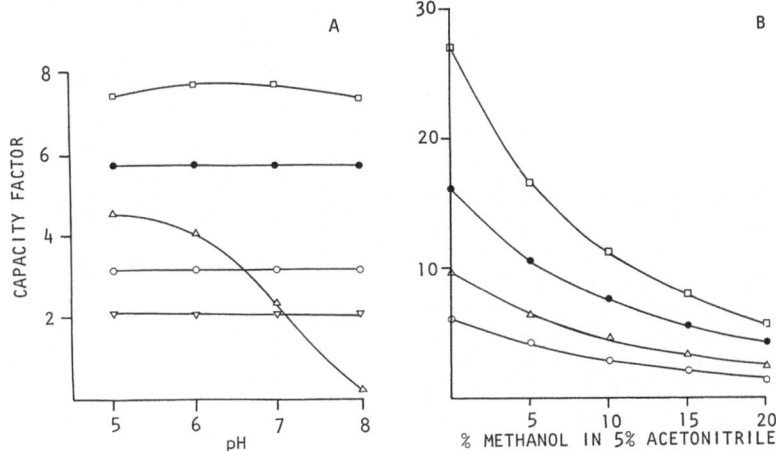

Fig. 3. HPLC capacity factors for antipyrine and its metabolites as influenced by: **A**, pH of eluent; **B**, methanol content of eluent. Symbols (for abbreviations, see text): AN, ● ; 4-OH-AN, □ ;   nor-AN, **A**: Δ; **B**, ∇;   3-OHMe-AN, **A**: ∇; **B**, o; 4'-OH-AN (**A** only): o. Column: as for Fig. 1. Eluent (*not* containing acetate buffer as in the procedure adopted): **A**, acetonitrile-methanol-0.05 M phosphate buffer of varying pH, 5:5:90 by vol.; **B**, as in **A**, but proportion of methanol to buffer (pH 5.0) varied. Flow-rate 1 ml/min.

Fig. 4. Chromatograms for anti-pyrine and its metabolites, isol-ated from urine as described in the text with hydrolysis done by **A**, acid, or **B**, enzyme treatment. Peaks: 1, 3-OHMe-AN; 2, nor-AN; 3, AN; 4, 4-OH-AN; 5, i.s. (AP). Column: as for Fig. 1. Eluent: acetonitrile-methanol-0.2 M pH 5.0 acetate buffer. Detection at 270 nm. For incubation conditions, see text   (metabisulphite added in **A** only).

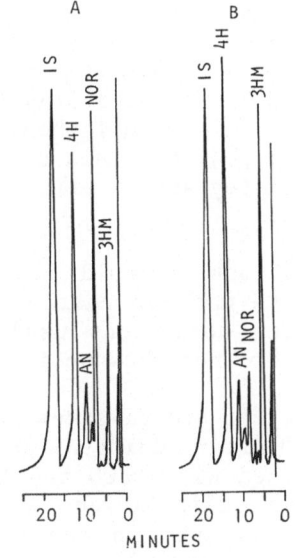

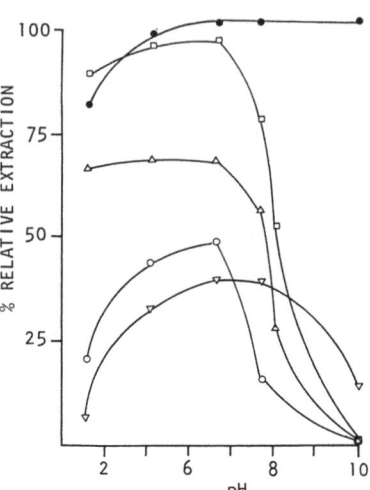

Fig. 5. Extraction of antipyrine and its metabolites into chloroform (5 vol.) at different pH values, as % of the AP extraction efficiency (= 98% at any pH). Phosphate buffers, 0.05 M; for other conditions, see text. Symbols (see text for abbreviations): AN, ●; 4-OH-AN, □; nor-AN, Δ; 3-OHMe-AN, ∇; 4'-OH-AN, o .

The instability observed for 4-OH-AN precludes the use of acid. It has been suggested [7] that the inclusion of sodium metabisulphite stabilizes the liberated 4-OH-AN. However, this approach was only partially successful in acid conditions, and caused enzyme inhibition (~50%) which an increase of enzyme concentration did not counteract. The best solution proved to be the use of a short incubation time (3 h) with high enzyme concentration (20,000 u./ml of urine).

Hydrolysis of 3-OHMe-AN conjugates was likewise complete under these enzymic conditions. The nor-AN conjugates proved less amenable, but could be cleaved rapidly by the acid procedure when sodium metabisulphite (2 mg/ml urine) was present to stabilize the product (as in Fig. 4A). An additional problem with nor-AN was adsorption onto glassware. 3-Amino-1-phenyl-2-pyrazolin-5-one has been suggested as a carrier [8]; but this nor-AN analogue gave serious interference in our chromatograms. Silanization of the glassware proved a better solution.

Although it could be resolved adequately in the HPLC system, 4,4'-OH-AN was too unstable to be analyzed successfully. With the extraction conditions used, degradation products formed during hydrolysis did not show up in the chromatograms.

Enzymic hydrolysates (pH 5) were extracted (5 vol. chloroform, unwashed) after adding AP as i.s. Cooled acid hydrolysates were first neutralized with NaOH and buffered to pH 5. Subsequent pre-chromatographic manipulations were as for AP metabolites. Any non-metabolized AN present in the urine shows up in either assay procedure, serving as a useful comparability check.

Table 1. Validation of the assays, as applied to urine (human; rat urine has also been used). There is linearity over the concentration ranges as tabulated, corresponding to levels found after administering a standard test dose. The relative recoveries are based on peak height ratios, using an i.s. (98% extraction efficiency) as stated in the text for each assay. Spiking was at mid-range level.

	Normal operating range, µg/ml	Relative recovery, %	Reproducibility (C.V.)
AP	1–10	101.1	3.9
Methylamino–AP	0.2–5	54.4	10.7
Amino–AN	5–10	76.1	7.2
N–Ac–amino–AN	20–40	89.5	4.1
AN	1–6	102.4	3.9
3–OHMe–AN	15–35	40.4	6.2
4–OH–AN	20–50	96.6	9.8
nor–AN	20–40	86.3	10.2

VALIDATION OF AP AND AN METABOLITE ASSAYS

Table 1 shows the ranges, as encountered in practice, over which the above assays have been used, together with recovery data and C.V. % values for the entire assay procedure performed on several consecutive days. Additional solvent extractions besides the single one actually performed would improve the lower recoveries.

Acknowledgements

For samples of certain metabolites, thanks are expressed to Prof. Schuppel of Braunschweig and to Prof. Yoshimura of Fukuoka. (Other authentic compounds were obtained from Aldrich, BDH or Hoechst.)

References

1. Stevenson, I.H. (1977) *Br. J. Clin. Pharmacol.* 4, 261–265.
2. Hildebrandt, A.G., Roots, I., Heinemeyer, G., Nigam, S. & Helge, H. (1979) in *The Induction of Drug Metabolism* (Estabrook, R.W. & Lindenlaub, E., eds.), Schattauer Verlag, Stuttgart, pp.615–627.
3. Lindgren, S., Collste, P., Norlander, B. & Sjoqvist, F. (1974) *Eur. J. Clin. Pharmacol.* 7, 381–385.
4. Chang, R.L., Wood, A.W., Dixon, W.R., Conney, A.H., Anderson, K.E., Eiseman, J. & Alvares, A.P. (1976) *Clin. Pharmacol. Ther.* 20, 219–226.
5. Meffin, P.J., Williams, R.L., Blaschke, T.F. & Rowland, M. (1977) *J. Pharm. Sci.* 66, 135–137.
6. Vesell, E.S., Passananti, G.T., Glenwright, P.A. & Dvorchik, B.H. (1975) *Clin. Pharmacol. Ther.* 18, 259–272.

7. Danhof, M., de Groot-van der Vis, E. & Breimer, D.D. (1979)
   Pharmacology 18, 210-223.
8. Hignite, C.E., Tschang, C., Huffman, D.H. & Azarnoff, D.L. (1978)
   Drug. Metab. Dispos. 6, 288-295.

#D-3

Analytical case history

IMMUNOASSAY OF BROMOCRIPTINE AND SPECIFICITY OF ANTIBODY: CRITERIA FOR CHOICE OF ANTISERUM AND MARKER COMPOUND

J. Rosenthaler, H. Munzer and R. Voges

Sandoz Ltd., Biopharmaceutical Department
4002 Basle, Switzerland

To develop an assay for bromocriptine (2-bromo-α-ergokryptine), different hapten compounds were used to produce various animal antisera. Since all these were predominantly specific to the tricyclic peptide moiety of bromocriptine, the affinity constant K was considered a sufficient criterion for selecting a high energy antibody. The K value for the favoured antiserum (#270979) was 0.9 x 10⁹ 1/M. With the favoured tracer ligand, a detection limit of 0.17 ng/ml was achieved for bromocriptine in equilibrium with its epimer bromocriptinine. Metabolites such as the 8'-hydroxy derivative show virtually no cross-reaction.

Bromocriptine, a synthetic 2-bromo derivative of the naturally occurring α-ergokryptine, in common with all ergot peptide alkaloids consists of two rigid ring systems, as shown: the one in the ergolene moiety and the other in the tricyclic peptide moiety *(shaded areas)*. They are connected by a flexible link, namely ring D as part of the ergolene moiety, and the central amide structure [1]; theoretically, free rotation is possible *(arrows)*. The main pharmacological action is inhibition of prolactin secretion [2]. Central dopamine agonist activity was later found [3], and led to new treatments for Parkinson's disease [4].

Assay in blood plasma calls for high sensitivity, most likely to be achieved by RIA. In competitive binding tests it is generally assumed that the binding characteristics of the labelled tracer compound are identical with those of the unlabelled ligand.

This requirement, critical for sensitivity, is fulfilled only if the tracer compound is labelled in such a way that the molecule is chemically identical with the ligand, as is normally attainable by radioactive labelling, substituting tritium for hydrogen or ^{14}C for carbon. Both these isotopes are weak β-emitters, and can be determined by liquid scintillation counting. Introduction of ^{14}C is in general the more difficult approach, and gives tracer compounds with low specific radioactivity; hence the tritium approach is commoner.

Sensitivity is also notably influenced by the affinity of the antiserum, a measure of the degree of binding between the antigen's determinant group and the antibody-combining sites. The affinity is defined as the equilibrium constant for a simple bimolecular reaction between a ligand containing only one epitopic group and an antibody containing binding sites of one class only [5]. Antisera as met with in practice are populations of antibodies containing mixtures having binding sites with different affinities. Yet a linear dose/response relation is frequently obtained within a certain range of concentrations, despite that heterogeneity.

An account is now given of trial of various tritium-labelled tracer compounds, and of antisera to pick the one which gives highest sensitivity and shows least affinity for the metabolites. Since bromcriptine epimerizes under some conditions, and also becomes concentrated on surfaces of various types, attempts to obviate these effects are described.

PREPARATION OF THE CONJUGATES FOR IMMUNIZATION

In view of the availability of 6-nor-6-carboxymethyl-9,10-dihydro-α-ergokryptine, a bromocriptine derivative of closely similar structure, it seemed obvious to take it as hapten and to couple it to the carrier protein, viz. thyroglobulin (#72797). Later on, the conditions of the Mannich reaction [6] between phenol and secondary amines in the presence of formaldehyde proved useful when coupling dihydro-α-ergokryptine (#010978) and eventually also bromcriptine (#270979) as hapten compounds to bovine serum albumin; either could be linked by their indole nitrogen to the accessible phenolic tyrosyl residues of the protein. For immunization rabbits and sheep were injected with 1 mg and 10 mg of these conjugates respectively, in several divided doses [7-9]. The titres of the three antisera were respectively 1:20,000, 1:200,000 and 1:30,000. Fig. 2 shows the conjugation positions, with the amount of conjugate bound to the protein for those conjugates that were soluble in aqueous solution.

AFFINITY CONSTANTS FOR THE ANTISERA

Response curves covering the concentration range 0.05-6.56 ng were produced for the 3 antisera. The equilibrium constants in the presence of tritiated dihydro-α-ergokryptine were calculated using

Fig. 1. Formation of conjugates for raising antisera:
1) #72797: 9,10-dihydro-α-
ergokryptine (R = H) attached
to thyroglobulin at N^6; 127
molecules of hapten/molecule
of protein.
2) #010978: as for 1), but to
bovine serum albumin at N^1;
conjugate insoluble.
3) #270979: bromocriptine (R
= Br, and 9,10 double bond)
attached as for 2); 21 mole-
cules/molecule of protein.

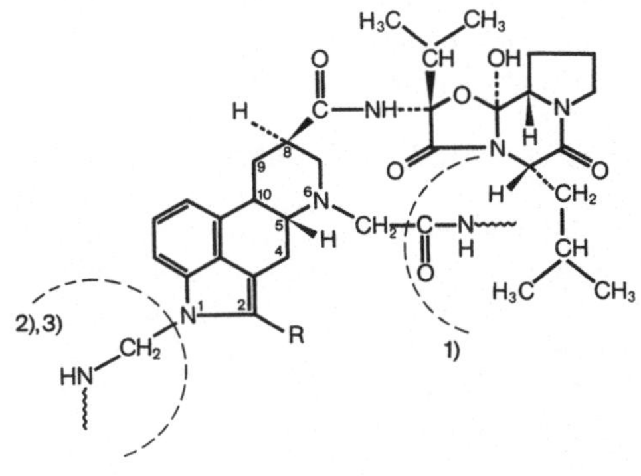

the Scatchard equation.   At equilibrium a higher energy antibody
will leave a lower concentration of free ligand in solution.   The
unit for K values is l/mol (it is expressed as concentration in the
final incubation mixture before adding charcoal).   With 9,10-dihydro-
$[13-^3H]$α-ergokryptine as tracer compound,   3 independent experiments
were performed with each antiserum;   essentially linear relations
were obtained in Scatchard plots, which is considered as evidence
for homogeneity of binding sites.   For the 3 antisera the respective
K values (with S.D.) were $0.34 \pm 0.01 \times 10^9$, $0.45 \pm 0.02 \times 10^9$,   and
$0.62 \pm 0.01 \times 10^9$.   Evidently in the series of antisera there is only
a small increase in affinity, i.e. the immunodeterminant part of the
ergot peptide alkaloid appears to be the peptide moiety.   Coupling
hapten to protein at N^1 yielded an antiserum of only slightly higher
affinity than coupling it at N^6.

On the other hand, when bromocriptine was used as hapten to
raise an antiserum, there was some increase in affinity.   This antiserum was then used in the RIA method development.   However, the
tritiated dihydro-α-ergokryptine still did not meet the conditions
required for very high sensitivity.   The two possible alternatives,
which we explored, were the radiolabelling with tritium either of
bromocriptine or of its 9,10-dihydro derivative.

A radiolabelled intermediate of high activity was obtained by
catalytic replacement of the Br atom in 13-bromo-9,10-dihydro-α-
ergokryptine using carrier-free tritium and 10% Pd on aluminium
oxide.   Bromination of the product with bromine in chloroform then
yielded the desired labelled ligand with the specific radioactivity
38 mCi/mg. Radiolabelling of bromocriptine with tritium was effected
by introducing tritium into the 6-methyl group.   Bromocriptine of
high specific radioactivity (35 mCi/mg) was obtained by reacting its
desmethyl derivative with tritiated methyl iodide.

The two radiolabelled ligands were evaluated in the presence of the bromcriptine antiserum #270979, in 5 independent experiments, and showed identical affinities. From Scatchard plots which again showed essentially linear relations, K was 0.92 ±0.04 × 10⁹ for ³H–bromocriptine and 0.93 ±0.04 × 10⁹ for ³H–dihydrobromocriptine.

CHOICE OF TRACER LIGAND

Since the two tracer compounds had virtually identical affinities and specific radioactivities, other criteria had to govern the choice between them. The decision was simplified by the observation that when tritiated bromocriptine which had been stored for a few weeks was used again in the RIA, non-specific binding accounted for as much as 25% of the radioactivity introduced, evidently due to instability of highly tritiated bromocriptine. Even when the labelled ligand was highly diluted, to 3×10^{-8} M, and the ethanolic solution was stored at 4°, non-specific binding increased with the duration of storage. The ³H–dihydro compound was therefore adopted as tracer.

SPECIFICITY OF THE VARIOUS ANTISERA

In RIA development, the specificity of the antiserum must be clearly defined, in view of possible cross-reaction with metabolites.

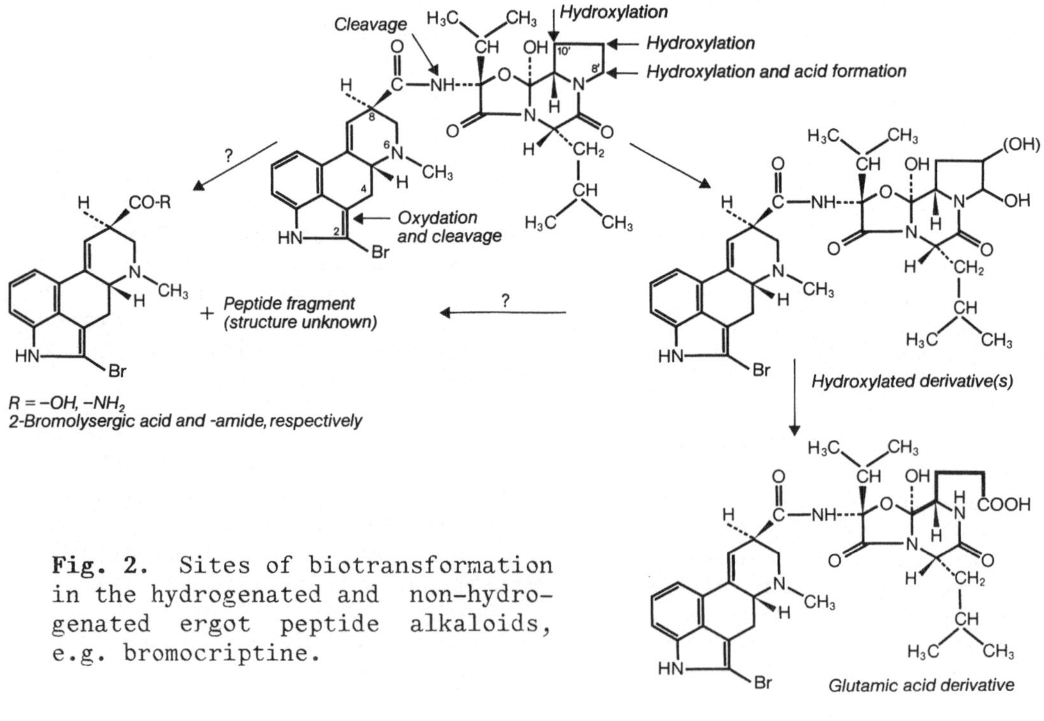

Fig. 2. Sites of biotransformation in the hydrogenated and non–hydrogenated ergot peptide alkaloids, e.g. bromocriptine.

Table 1. Inhibitory 'dose' (ID$_{50}$) of bromocriptine, bromcriptinine, and a metabolite of closely related structure, at 50% binding to 3 antisera (see text). (In parentheses: no. of experiments.)

Anti-serum	Compound	ID$_{50}$ ±S.D., ng	Ratio
#72797	Bromocriptine	3.11 ±0.04 (3)	[1.0]
	Bromocriptinine	22.8 ±2.4 (3)	7.3
	8'-hydroxyl derivative (αC-8)	333.4 (1)	107
#010978	Bromocriptine	3.15 ±0.15 (3)	[1.0]
	Bromocriptinine	23.0 ±0.8 (3)	7.3
	8'-hydroxyl derivative (αC-8)	559.8 (1)	178
#270979	Bromocriptine	1.87 ±0.02 (3)	[1.0]
	Bromocriptinine	17.7 ±0.9 (3)	9.5
	8'-hydroxyl derivative (αC-8)	342.6 (1)	183

The metabolism of bromocriptine is especially difficult to study, as enteral absorption is incomplete and excretion occurs mainly with the bile via the liver, little being excreted in the urine. There are several metabolites, with possible epimers. Bromocriptine is metabolized similarly to 9,10-dihydro-β-ergokryptine (G. Maurer et al., to be published). Generally, it appears, ergot peptide alkaloids are hydroxylated first at the 8'-position (Fig. 2). Further hydroxylation takes place at the 9'-position, and the products may be in the normal β-form or the isomeric α-form with respect to the configuration at the chiral C-8 atom. Ring-opening at C-8' produces a carboxylic acid, structurally a glutamic acid derivative of the ergot peptide molecule. The next step is probably hydrolytic cleavage of the central amide bond, producing a lysergic acid moiety and a peptide moiety, viz. 2-bromo-lysergic acid or its amide and the corresponding iso-compounds (for amplification, see ref. [10]). The fate of the tricyclic peptide residue is still uncertain, but as will be shown later this is unimportant in the study of cross-reactions.

Some metabolites have been obtained both by synthesis and by isolation from rat urine. Table 1 shows similar specificities for the 3 antisera. None of them seems to discriminate adequately between bromocriptine and its isomer. On the other hand, with the 8'-hydroxylated metabolite of the α-form, the only one available at that time, 50% binding to antiserum is attained only when it is present in about 180-fold excess. At >5,000 ng concentration, other metabolites such as both epimers of the glutamic acid analogue, 2-bromolysergic acid amide and 2-bromolysergic acid showed no interaction with any of the antisera; Fig. 3 shows the structures of the various compounds tested.

The antisera which were produced by means of conjugates 1), 2) and·3), i.d. ◆72797, ◆010978, and ◆270979 were tested for cross-reaction with:

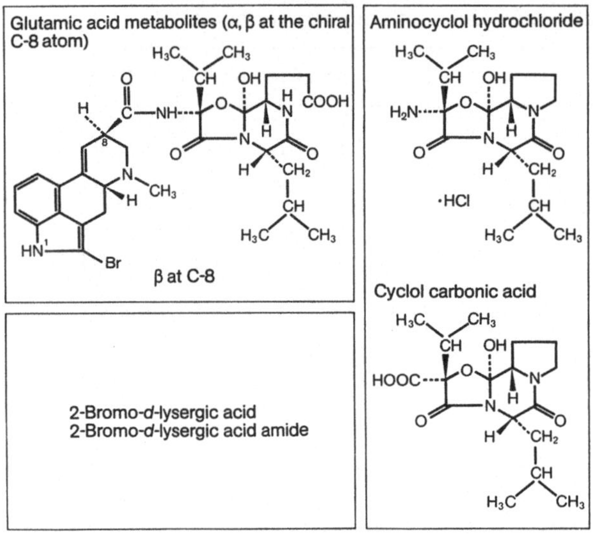

Glutamic acid metabolites (α, β at the chiral C-8 atom)

β at C-8

2-Bromo-d-lysergic acid
2-Bromo-d-lysergic acid amide

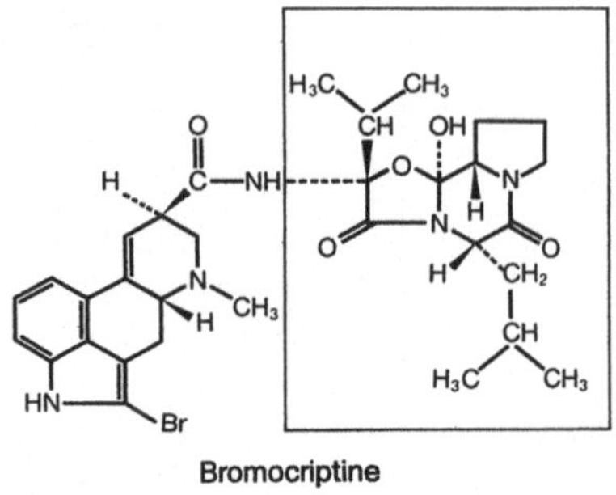

Aminocyclol hydrochloride

·HCl

Cyclol carbonic acid

Bromocriptine

Fig. 3. Some compounds – including two closely related to the drug's tricyclic peptide moiety *(see right-hand panel)* – that were tested for cross-reaction, using the three antisera.

RADIOIMMUNOASSAY FOR BROMOCRIPTINE

The antiserum employed in developing an RIA procedure was produced from a conjugate based on bromocriptine as hapten. The selected tracer compound is specified in Scheme 1, which outlines the assay procedure. With 0.5 ml of plasma the detection limit is 0.17 ng/ml. The use of such a large aliquot has the disadvantage, especially in the RIA of bromocriptine, that matrix effects are observed. A target for future work, which may allow the volume of the unknown sample to be reduced, is to achieve a still higher radiolabel in the tracer molecule selected.

In applying the method (Scheme 1), particular emphasis was laid on the recovery of unchanged compound in samples specially prepared with plasma for quality control. Table 2 shows the results for 3 concentrations, distributed over the range covering the dose/response curve.

Because bromocriptine becomes concentrated onto glass and plastic surfaces, its aqueous solutions cannot be diluted without losses. This must be borne in mind when serial dilution is done in preparing the standard curve. In this process bromocriptine is lost by adsorption, and the actual concentrations obtained are lower than those expected, to an extent corresponding to the amount adsorbed. The only feasible means of verifying this was to carry out dilution experiments on ^{3}H-bromocriptine of high specific radioactivity in two buffer systems frequently used in the RIA. Table 3 shows that bromocriptine solutions may be diluted with blood plasma or with aqueous

The following additions (ml) to the specified tubes are made in turn:	TA	NSB	Bo	Std	Unk
Buffer: 0.1 M citrate containing 0.5% sodium azide, used at pH 6.0 ±0.1:-	1.1	0.6	0.5	0.5	0.5
Plasma: charcoal-treated human plasma, except for **Unk** where an unknown plasma sample is used:-	0.5	0.5	0.5	0.4	0.5
Standard: 2-bromo-α-ergokryptine methane-sulphonate (range: 0.08-10 p-mol):-	-	-	-	0.1	-
Tracer: 2-bromo-9,10 dihydro-(13-^{3}H)α-ergokryptine (base) in 1% Solulan:-	0.1	0.1	0.1	0.1	0.1

The tube contents are mixed; then add:

Antiserum: anti-2-bromo-α-ergokryptine:-		-	-	0.1	0.1	0.1

Mix again, and incubate 1 h at 37° and 1 h at 4°; then add as ice-cold suspension:

Charcoal coated with blood plasma:-		-	0.5	0.5	0.5	0.5

Incubate for 10 min at 4°, centrifuge for 5 min at 4° (1,200 g), decant, add 10 ml scintillator fluid to whole supernatant, and count.

Scheme 1. RIA protocol for bromcriptine in human plasma. Tubes: **TA**, total radioactivity; **NSB**, non-specific binding (radioactivity not adsorbed by charcoal in the absence of antibody); **Bo**, bound labelled ligand with no unlabelled ligand present; **Std**, standard; **Unk**, unknown, i.e. plasma containing ligand to be assayed.

Table 2. Testing precision and accuracy by analyzing quality control samples of bromocriptine. The samples at 3 levels (ng/0.5 ml) in human plasma were analyzed, in duplicate, 10 times using 10 different RIA kits of the same batch. Estimates of variation represent relative standard deviation (= coefficient of variation, C.V.). The accuracy values are given as % deviation from reference level (= 100).

	Bromocriptine methane sulphonate, reference level		
	6.56	1.31	0.07
Mean value found, $\bar{x}$ (& S.D.)	6.85 (+ 4.4%)	1.30 (−0.8%)	0.08 (+14.3%)
Within-assay variation, s_w [& C.V.]	0.29 [4.3%]	0.11 [8.3%]	0.014 [17.9%]
Between-assay variation, $s_{\bar{x}}$ [% C.V.]	0.41 [6.0%]	0.12 [8.9%]	0.016 [20.6%]

Table 3. Dilution of tritiated bromocriptine with different media. Arithmetic–scale dilutions were performed. The phosphate–buffered saline (PBS) contained 10 mM phosphate in 0.1 M saline, pH 7.4. The nos. in parentheses () denote % recovery compared to accepted reference level. The drug was used in methanesulphonate form.

Ref. level, pg/ml	70% aqueous ethanol	PBS	PBS with 0.1% bovine serum albumin	Human plasma, charcoal–tr'd
1,200	1,200 (100)	1,131 (94.2)	1,067 (88.9)	1,182 (98.5)
600	604 (100.7)	134 (22.4)	333 (55.6)	588 (98.1)
300	300 (100)	34.1 (11.3)	126 (42.3)	293 (97.5)
150	151 (100.4)	15.2 (10.1)	55.5 (37.0)	150 (99.9)
75	74.2 (98.9)	7.2 (9.6)	22 (29.4)	73.1 (97.5)
37.5	37.4 (99.9)	2.5 (6.8)	9.6 (25.7)	36.3 (96.8)
18.75	18.1 (96.3)	1.0 (5.3)	4.1 (22.2)	18.1 (96.3)

ethanol, whereas phosphate–buffered saline is unsuitable even if 0.1% bovine serum albumin is present.

Another problem in the RIA of bromocriptine is the possibility of epimerization at two chiral atoms, i.e. C-8 and C-2'. This may give rise to an equilibrium between 4 molecular species: bromocriptine ⇄ bromocriptinine, and aci–bromocriptine ⇄ aci–bromocriptinine. It is thought that the latter reaction is rather unlikely to occur in body fluids, despite the elevated temperature. However, epimerization at the chiral C-8 atom does occur; the molecule isomerizes in neutral or especially alkaline solution. Table 4 indicates the extent of isomerization as shown by HPTLC. Evidently equilibrium is reached very quickly in blood plasma. It remains unchanged during the RIA incubation period, the concentration of the epimeric form being about 20%. Thus it is hard to prove that the antiserum used is specific only for the normal form of the alkaloid. When the α–isomer is incubated in plasma under the same conditions, 13% of the β–isomer is formed. It is conceivable that in raising the antiserum the bromocriptine present as hapten in the conjugate may have been partly isomerized, and antibodies specific for the α–isomer thus produced. Since both epimers are formed under incubation conditions, whether the α–form or the β–form is used as starting material, the specificity of the antibody cannot be definitely delineated.

The important feature of the RIA for bromocriptine is that the equilibrium present in blood plasma is unchanged under incubation conditions, and therefore the concentration of bromocriptine in the body fluids can be reproducibly measured.

Table 4. Stability of bromocriptine and of its epimer. The values represent the % of either compound formed by inversion at the chiral C-8 when dissolving. Each solution contained 29.15 µg/ml. Citrate/plasma signifies 0.7 ml of 0.1 M pH 6.2 citrate + 0.5 ml plasma; kept 1 h at 37° and 1 h at 4°. PBS/plasma signifies 0.7 ml PBS (see Table 3) + 0.5 ml plasma; kept 1 h at 37°. Following these incubations, some samples were kept at 4° for 16 h. The incubation of the plasma without buffer was for 1 h at room temperature. HPTLC was on silica gel (F254) with CH_2Cl_2-rich solvent + NH_3; finally irradiate at 307 nm.

	70% aq. ethanol	Citrate	PBS	Plasma
Bromocriptine				
Immediately after dissolution	<0.5	<0.5	<0.5	<0.5
After incubation	–	20.1	16	20.6
After 16 h at 4°	–	27	20.5	–
Bromocriptinine				
Immediately after dissolution	4.1	5.5	6	9
After incubation	–	9.8	10.3	13.5
After 16 h at 4°	–	10.7	9.5	–

References

1. Weber, H.P.(1980) in *Ergot Compounds and Brain Function* [Vol. 23, *Adv. Biochem. Psychopharmacol.*] (Goldstein, M., Calne, D.B., Lieberman, A. & Thorner, M.O., eds.), Raven Press, New York, pp. 25–34.
2. Flückiger, E. & Wagner, H.-R. (1968) *Experientia 24*, 1130.
3. Hökfelt, T. & Fuxe, K. (1972) in *Brain-Endocrine Interaction* (Knigge, K.M., Scott, D.E. & Weindl, A., eds.), Karger, Basel, pp. 181–223.
4. Calne, D.B., Teychenne, P.F., Claveria, L. E., Eastman, R., Greenacre, J.K. & Petrie, A. (1974) *Br. Med. J. 4*, 442–444.
5. Yalow, R.S. & Berson, S.A. (1971) in *Competitive Protein Binding Assays* (Odell, W.D. & Daughaday, W.H., eds.), Lippincott, Philadelphia, pp. 1–21.
6. Levy, A., Kawashima, K. & Spector, S. (1976) *Life Sci. 19*, 1421–1429.
7. Rosenthaler, J. & Munzer, H. (1976) *Experientia 32*, 234–235.
8. Rosenthaler, J., Nimmerfall, F., Sigrist, R. & Munzer, H. (1977) *Eur. J. Biochem. 80*, 603–609.
9. Rosenthaler, J., Pacha, W.L., Koeberle, S. & Beveridge, T. (1980) in *Phenothiazines and Structurally Related Drugs: Basic and Clinical Studies* (Usdin, E., Eckert, H. & Forrest, I.S., eds.) Elsevier/North Holland, New York, pp. 121–124.
10. Eckert, H., Kiechel, J.R., Rosenthaler, J., Schmidt, R. & Schreier, E. (1978) in *Ergot Alkaloids and Related Compounds*, Vol. 49, *Handbook of Experimental Pharmacology* (Berde, B. & Schild, H.O., eds.) Springer-Verlag, Berlin, pp. 780–782.

#D-4

Analytical case history

DETERMINATION OF 3-PHENOXY-*N*-METHYLMORPHINAN
AND METABOLITES IN PLASMA

J.A.F. de Silva, J. Pao and M.A. Brooks

Department of Pharmacokinetics and Biopharmaceutics
Hoffmann-La Roche Inc.
Nutley, NJ 07110, U.S.A.

Requirement /metabolism
: *A sensitive assay for the drug and 4 Phase I meta-bolites, including levorphanol formed by cleavage of the ether linkage (an active metabolite).*

End-step
: *(1) GC-MS with positive chemical ionization (+ve CI), or (2) GC-AFID (N-P detector, NPD); also tried (3): HPLC-UV.*

Sample preparation
: *Heptane extract from 1 ml plasma at pH 11 dried down, preceded in the case of (2) by back-extraction into acid and re-extraction.*

Comments
: *(1) Satisfactory for the drug and, after derivatization, notably successful for levorphanol. (2) Of some use for the drug; but lengthy, and sensitivity only fair (~5-10 ng/ml). (3) Poor sensitivity because of poor UV absorption.*

The drug (**I**, Fig. 1) is one of the phenoxymorphinan analogues synthesized by Mohacsi [1] and is under development as a non-narcotic analgesic. It is extensively metabolized in the rat [2] and in the dog [3] by *N*-demethylation to form the nor-analogue (**I-A**), and by *p*-hydroxylation of the 3-phenoxy ring to yield **I-B** and cleavage of the ether linkage to yield the morphinan **I-C** and its *N*-demethylation produces nor-levorphanol (**I-D**) (Fig. 1). The presence of four metabolites besides the drug necessitated a strategy for the development of sensitive and specific assays for their quantitation. Of the approaches tried, with dog plasma during pre-clinical drug development, GC-MS (+ve CI) proved the most satisfactory.

Phenoxymorphinans:

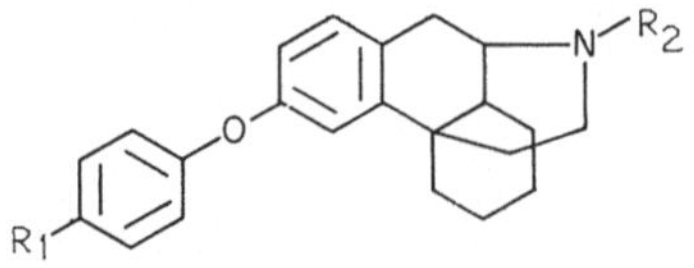

	R_1	R_2	
I	–	CH_3	(–)-3-Phenoxy-N-methylmorphinan
I-A	–	H	(–)-3-Phenoxymorphinan
I-B	HO	CH_3	(–)-3-(p-Hydroxy)phenoxy-N-methylmorphinan
II	CH_3	CH_3	(–)-3-(p-Methyl)phenoxy-N-methylmorphinan
III	–	CH_3	(–)-3-Pentadeutero-phenoxy-N-methylmorphinan

Morphinans:

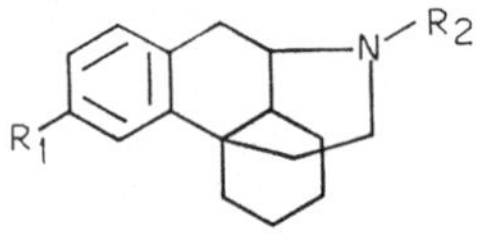

	R_1	R_2	
I-C	HO	CH_3	(–)-3-Hydroxy-N-methylmorphinan (Levorphanol)
I-D	HO	H	(–)-3-Hydroxymorphinan (Norlevorphanol)
IV	HO	$CH_2CH\text{-}CH_3$	(–)-3-Hydroxy-N-allylmorphinan (Levallorphan)
V	CH_3O	CH_3	(+)-3-Methoxy-N-methylmorphinan (Dextromethorphan)

Fig 1. Chemical structures of phenoxy-N-methylmorphinan, its major metabolites and related analogues. V = i.s. for HPLC assay; II = i.s. for the GC-AFID(NPD) assay; III = i.s. for GC-MS(+ve CI) assay of I, and IV = i.s. for I-C.

Each plasma sample (1 ml), after addition of 2 ml pH 11 1 M phosphate buffer and 2 ml dist. water, was extracted with n-heptane (10 ml; done twice in the GC-AFID approach) in a 50 ml centrifuge tube with a PTFE stopper. The final extract, after the indicated clean-up in the case of GC-AFID, was dried down, and the residue re-dissolved as stated below for each approach. Appropriate standards, as specified, were spiked in at the outset; I was of pharmaceutical grade purity (>99%), and III of purity >98.6% (synthesized by W. Burger and A. Liebman, Radiochemical Synthesis Group, Chemical Research Division, Hoffmann-La Roche Inc.).

HPLC APPROACH (FOR I, I-A, I-B)

Trial runs as in Fig. 2 established that 10 ng/ml of I or 30 ng/ml of I-A was measurable in plasma; but the sensitivity limits were inadequate for I-B and I-C (200 and 500 ng/ml) due to their poor UV absorbance, and were not significantly improved by fluorimetric detection (280/315 nm). Whilst the HPLC-UV approach did show the presence of I and its phenoxy metabolites I-A and I-B in the dog after high dosage, 100 mg/kg (Fig.2, D), its limitations warranted trial of GC-AFID as offering sensitivity and specificity.

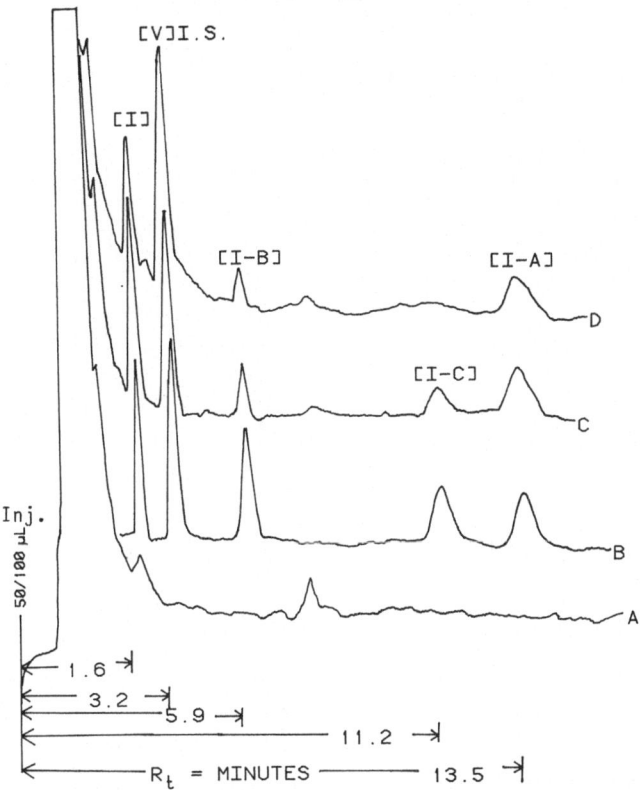

Fig. 2. HPLC runs on: **A**, extract from control dog plasma; **B**, authentic standards; **C**, authentic standards recovered from plasma equivalent to 25 ng each of **I**, **I-A** and **I-B**, 250 ng of **I-C** and 400 ng of **V** (i.s.);   **D**, dog plasma drug/metabolite profile 6 h after oral dosing. In Figs. 2-4, Inj denotes the injection point.
Column: Partisil (Whatman), similar to the silica columnn used in a flurazepam study (this vol., #D-1). Mobile phase: dichloromethane
–methanol–conc. NH₃, 90:9.4:0.6 by vol.; 2.0 ml/min. Detection: 254 nm.
Injection: 50 μl from the 100 μl of dichloromethane:methanol (9:1 by vol.) used to dissolve the residue from a *n*-heptane extract of plasma.
[R_t ≡ other authors' t_R.–*Ed*.]

GC–AFID (NPD) APPROACH

Whilst the metabolites **I-A** and **I-C** could be analyzed besides **I** in a single run, the **I-C** peak showed tailing which would have called for derivatization of the phenolic OH group as the remedy. Since **I-A**, **I-B**, **I-C** and **I-D** were found by ³H₁-labelled **I** administration to be only minor metabolites in dog plasma [4], the aim was amended to the specific determination of **I** for pre-clinical studies in the dog. Fig. 3 shows typical chromatograms.   The overall recovery of **I** in the range 5–100 ng/ml of plasma, with 300 ng of **II** as i.s., was 65±5.0% (S.D.).   The sensitivity limit for quantitation was 5–10 ng/ml.

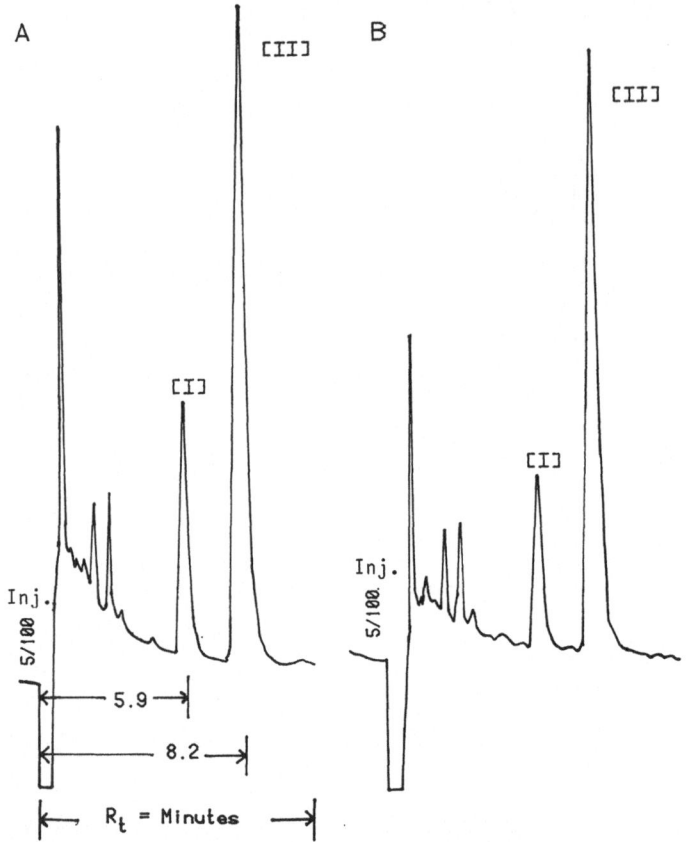

Fig. 3. GC-AFID (NPD) runs, on: **A**, compounds spiked into dog control plasma, viz. 75 ng/ml of the parent drug **I** and 300 ng/ml of the analogue **II** as i.s.; **B**, dog plasma 1 h after a 10 mg/kg oral dose (**I** reckoned to be 57 ng/ml).
Column: 1.22 m × 2 mm i.d.; 2% OV-17 on 100/120 Gas Chrom Q, in a GC apparatus (Hewlett-Packard, 5170A) equipped with an NPD (16 V DC). Injector at 250°, column at 230°, detector at 300°. Flow rates, ml/min: He, 1.4; H_2, 2.5; air, 4.0.
Injection: 5 µl from the methanolic solution (100 µl) of the residue obtained by back-extracting into acid, reextracting and drying down.

In a dog given **I** (10 mg/kg) orally as its hydrochloride in a hand-packed gelatin capsule, the plasma concentration declined from its peak at 10 min (175 ng/ml) to ~7 ng/ml at 11 h ($t_{1/2}\beta$ ~7.4 h). With a lower dose the sensitivity could not have coped with the rapid metabolism and elimination of **I**. The assay was also time-consuming. The method of choice, as illustrated in Fig. 4, was GC-MS for which the sample did not need extensive 'clean-up'. As will now be considered, the approach was notably effective for assaying **I** *per se* in plasma.

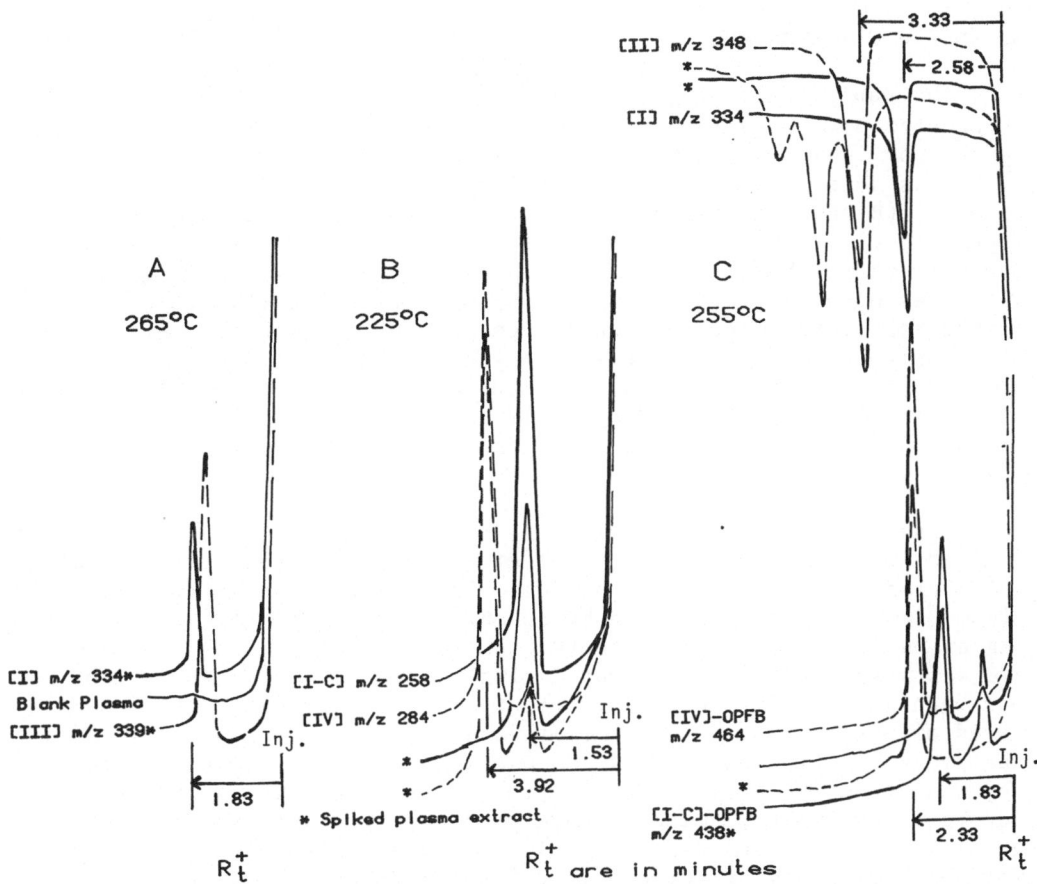

Fig. 4. Ion chromatograms from GC–MS(+ve CI) runs, on: **A**: parent drug **I** and its i.s. **III**; **B**, the metabolite **I-C** and its i.s. **IV**, determined without derivatization; **C**, SIM analysis of the penta-fluorobenzyl ethers of **I** at m/z = 334, **II** as i.s. at m/z = 348, the metabolite **I-C** at m/z = 438, and its i.s. **IV** at m/z = 464. Curves marked bl and sp were obtained with extracts from blank and spiked plasma respectively, as distinct from authentic standard for which a corresponding line (unbroken or broken) is used.

Column: either 1.22 m x 2 mm i.d. (pre-siliconized borosilicate glass) packed with 3% OV–17 on 120/140 mesh Gas Chrom Q, or, for C, 1.83 m x 2 mm i.d. packed with 3% OV–1 on 100/120 mesh Gas Chrom Q. The stated temperatures refer to the oven; for **A** the port was at 275° and the interface at 215°. (For apparatus, see text.) Reagent gas: ammonia, via the direct insertion probe inlet; carrier gas, methane at 10 psig head pressure; methane ion source pressure 0.5 Torr, and total ion source pressure 0.6 Torr with ammonia added.

MS operation: ion energy source, repeller and collector at +10 eV lens at –40 V, electron energy source –200 V, filament emission at 1.1 mA; electron multiplier at –2000 V, continuous dynode electron multiplier at –2500 V. See text for additional particulars.

GC–MS(+ve CI) APPROACH

The equipment (Finnigan, 9500 & 3200) was a splitless–injector GC coupled via a glass capillary restrictor to the MS, modified 'in–house' for both +ve CI [5] and –ve CI [6].   Selected ion monitoring (SIM) was done with a peak monitor (Promim, Finnigan) and a 4–channel recorder (Rikadenki, KA–41).   GC and MS conditions are given in the legend to Fig. 4.   The [MH]+ ions for I and its i.s. III at m/z 334 & 339 respectively were eluted at 1.83 min (Fig. 4, **A**).   The divert valve was open to vent out the first 30 s of column effluent, and then closed for it to enter the ion source and the filament current turned on.   The peak monitor channels were run at 10^8 V/amp gain, 0.1 s dwell time and filter at 0.5 Hz. Adsorption losses in the assay were minimized by initially adding 200 ng of **IV** as carrier to the 1 ml of plasma, besides 25 ng of **III**.   From the final chloroform solution (100 µl), 5–10 µl was taken for GC–MS.

Assays in the spiking range 1–50 ng/ml gave a C.V. of 17.8% at 2.5 and 5.6% at 25 ng/ml; 2.5 was effectively the sensitivity limit. Recoveries were in the range 65–70%.   The approach was effective with plasma samples not measurable by GC–AFID, e.g. from a dog given 10 mg /kg of encapsulated I.tartrate (9.6 ng/ml peak). The active metabolite I–C was also detectable without derivatization (Fig. 4, B; peak at 10h). It has been assayed in cancer patients by RIA [7]. Derivatization as now tried offers promise of increased sensitivity and reliability: extractive alkylation as applied to phenolic compounds such as morphine [8] was effective (Fig. 4, **C**), whereas direct reaction with PFB–bromide (diisopropylethylamine as catalyst) gave low yields. The ethers as obtained from the dried–down dichloromethane medium after TBAHS extraction and washing at acid pH gave [MH]$^+$ ions as shown.   The linearity and reproducibility (r = 0.9971) were good, but not the yield, although this has been improved in a modified version of the **I–C** assay [9].

Acknowledgements

The authors thank W.A. Garland and B.J. Miwa for assistance with MS analysis, and Miss N. Bekesewycz for preparing this publication text.

References

1. Mohacsi, E. (1978 & 1981) *U.S. Patents* 4,113,729 & 4,247,697 resp.
2. Kamm, J.J., Szuna, A. & Mohacsi, E. (1979) *Pharmacologist 21*, 173 (#156).
3. Leinweber, F.-J., Szuna, A.J., Williams, T.H., Sasso, G.J. &   De-Barbieri, B.A. (1981) *Drug Metab. Dispos. 9*, 284–291.
4. Leinweber, F.-J., Loh, A.C., Carbone, J.J. & Patel, I. H. (1980) *Unpublished data on file*, Hoffmann–La Roche Inc., Nutley, N.J.
5. Min, B.H. & Garland, W.A. (1977) *J. Chromatog. 139*, 121–133.
6. Garland, W.A. & Min, B.H. (1979) *J. Chromatog. 172*, 279–286.
7. Dixon, W.R., Crews, T., Mohacsi, E., Inturrisi, C. & Foley, K. (1981) *Res. Comm. Chem. Path. Pharmacol. 32*, 545–548.
8. Cole, W.J., Parkhouse, J. & Yousef, Y.Y. (1977) *J. Chromatog. 136*, 409–416.
9. Min, B.H., Garland, W.A. & Pao, J. (1982) Submitted for publication.

#D-5

Analytical case histories

ASSAYS WHERE HIGH SENSITIVITY WAS ACHIEVED, ESPECIALLY BY HPTLC WITH AUTOMATIC SPOTTING AND COLOUR REACTIONS

W. Ritter

Bayer AG, Department of Pharmacokinetics
D-56 Wuppertal 1, West Germany

The general account (#A-6, this vol.) of the scope and the equipment choice for thin-layer chromatography, particularly the high-performance mode (HPTLC), is now illustrated by four accounts of method development which, even where done before HPLC became widespread, has argued against HPLC or GC. Colour reactions as devised or adapted were crucial, as was the choice of a spotting instrument (#A-6, Appendix). The analytes and the approaches were as follows.-

Requirement	*Procedures for the determination, down to a few ng /ml plasma, of (1) muzolimine, (2) an indazole-type analgesic, in ester form and as an active metabolite, (3) a pyrazoline (nafazatrom) and metabolites, (4) imidazole antibiotics.*
End-step	*For (1), conventional TLC sufficed; for (2)-(4), HPTLC. Post-chromatographic derivatization to give coloured spots, then densitometry. Vol. applied in HPTLC: 40 µl.*
Sample preparation	*For (1), a dichloromethane extract (0.5 ml/ml plasma) is directly spotted; similarly (0.25 ml) for (2), urine or plasma. (3): 0.4 ml chloroform. (4): 2 ml hexane twice; dry down (N$_2$, 40°), redissolve in 0.4 ml chloroform.*
Comments	*For (1), account has to be taken of in vivo instability and of analytical losses with small amounts. For (2) the assay has shown abundant metabolite but little parent compound in plasma and urine. For (3), studied only in animal experiments, analytical losses tended to occur, precluding UV densitometry without colour generation (feasible if acid-pH HPTLC performed, but interferences arise).* *For (4), colour generation by picryl chloride is an effective final step that has wide applicability.*

(1) Muzolimine (BAY g 2821; Edrul^R)

This drug, 3-amino-1-(3,4-dichloro-α-methylbenzyl)pyrazolin-5-one, is a diuretic and antihypertensive. It is unstable in body fluids and on silica-gel TLC plates (which mitigates against assay by quantitative TLC), and it 'disappears' during clean-up through decomposition and/or adsorption onto glass (silanization or use of quartz vials did not help). GC was precluded because of irreproducible thermal decomposition, and HPLC was not available at that time (1974).

The detection limit on TLC plates, quantitated densitometrically at 225 nm without derivatization, was about 100 ng/ml, inadequate for pharmacokinetic studies: as was found later, 0.3 mg/kg in man - the lowest effective dose - produces peak plasma levels of about 200 ng/ml [1].

The solution of the problem lay in the above-mentioned single-step extraction and direct spotting, thus avoiding decomposition and /or adsorption onto glass associated with the usual procedure of evaporating down and re-constituting in a small volume of organic solvent. After development of the chromatogram, derivatization was performed by spraying with a solution of 4-dimethylaminocinnamaldehyde and heating in an oven at 100° for several minutes, yielding blue spots on a white background with maximal absorption at 610 nm [2].*

Thereby a detection limit of 1 ng/ml is achievable. The stained spots are stable for weeks. There is good precision because of the direct spotting of the extract from plasma or urine. Even in 1981, the detection limit for an HPLC assay was 100 ng/ml!

The assay as described enabled plasma levels to be determined in man during 50 h following a single oral dose of 0.3 mg/kg [1]. The assay was later modified for quantitative HPTLC: dipping instead of spraying produced an even more homogeneous background, and use of the Autospotter (DESAGA) enabled the standard solutions of the pure substance to be replaced by spiked body-fluid samples. However, although the detection limit for the pure substance was 50 pg/spot on HPTLC plates, sensitivity per ml of body fluid remained at the 1 ng/ml level (actually 100 pg/spot), because lower amounts 'disappeared' during extraction and spotting (Merck silicagel 60 plates, 10 × 20 cm).

---

* The composition of colour reagents is given in *Appendix below.−Ed.*

It was assumed that the product of reaction between muzolimine and 4-dimethylaminocinnamaldehyde is a Schiff's base [2]. Later it was found that there is also an aldol condensation at position 4 of the pyrazoline ring [cf. (2) & (3) below]. The structure of the blue-coloured derivative is therefore:

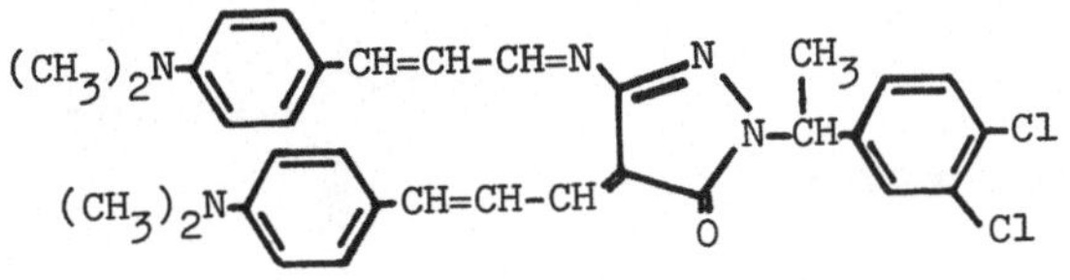

(2) BAY f 1936

The case history for this analgesic, 3-amino-5-trifluoromethyl-indazole-1-carboxylic acid ethyl ester, illustrates, in respect of a metabolite, rapid identification and rapid development of an assay.

With the experience of reacting muzolimine with an aldehyde [(1) above], an HPTLC assay was devised immediately without trying GC or HPLC. The developed HPTLC plate was dipped into a solution of the aldehyde and heated at 100° to yield red-violet spots (max. at 540 nm). The detection limit was 10 ng/ml for plasma put through the extraction procedure (described in the précis; routinely shaking, 10 min, and centrifuge).

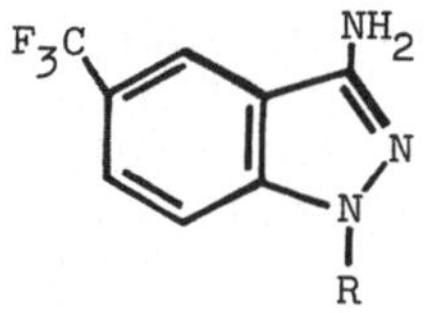

BAY f 1936: R = $COOC_2H_5$

BAY h 1127: R = H

However, following oral administration to experimental animals the concentrations of BAY f 1936 were very low, but there appeared on the chromatogram another spot having the same colour, but more polar, apparently a metabolite still having the 3-amino group. By comparison with a known compound, the metabolite was identified within hours as 3-amino-5-trifluoromethyl-1H-indazole, BAY h 1127 (SKF, U.S. patent no. 3133081 of 30 Nov. 1961). In man, following oral administration, only traces of the parent compound were found in plasma and urine, but high concentrations of BAY h 1127 as the main metabolite, obviously generated during first pass through the liver, i.e. BAY f 1936 serves as a pro-drug for BAY h 1127. Assay of the metabolite was done by the method developed for the parent drug, without any change or loss of time except that the synthesized metabolite was substituted for the parent drug as standard.

Later it was found that primary amines in general can be

quantitated by high-performance thin-layer densitometry, after post-chromatographic derivatization with 4-dimethylaminocinnamaldehyde to the corresponding Schiff's base.

(3) Nafazatrom (BAY g 6575)

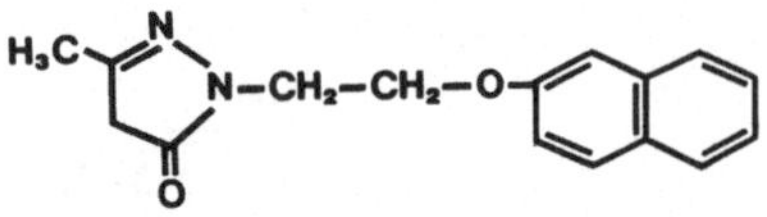

This antithrombotic [3, 4] and antimetastatic drug, 3-methyl-1[2-(2-naphthyloxy)-ethyl]-2-pyrazolin-5-one, is very unstable in body fluids and on TLC plates. Like muzolimine, nafazatrom 'disappeared' during clean-up procedures. GC was precluded because nafazatrom is not volatile, and HPLC was at that time not yet available to us (the TLC assay was developed in 1975).

Nafazatrom could be measured densitometrically in the UV at 225 nm, but the drug was too unstable and decomposed on the TLC plate during scanning. Under acidic conditions, i.e. after running the TLC plate with a solvent system containing acetic acid, the drug spots were more stable. The 10-20 ng/ml detection limit was sufficient; but the UV measurement was not specific enough because of interference by endogenous compounds that could not be separated chromatographically.

Finally a post-chromatographic derivatization reaction was found, viz. the aldol condensation of nafazatrom with 4-dimethyl-aminobenzaldehyde. The TLC plate was sprayed with a solution of the aldehyde, and heated at 100°, to yield orange spots on a white background. The stained spots are stable for weeks and are scanned at 490 nm. Nafazatrom analogues with one or two substituents in position 4 of the pyrazoline ring did not give the colour reaction with the aldehyde, which is evidence that the structure is indeed thus:

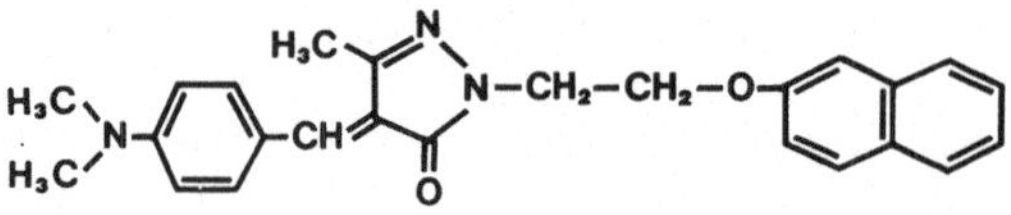

Metabolites with an unchanged position 4 of the pyrazoline ring are also detected by this reaction. Following oral administration, 3 metabolites were found in human urine (Fig. 1) (and 3 more in urine from a patient given repeated doses). The metabolites are assayed together with the parent drug during a single scan.

For quantitative HPTLC by this assay, spraying was replaced by dipping as a slight modification, and eventual improvement came from use of the Autospotter. As 20 samples could then be spotted simul-

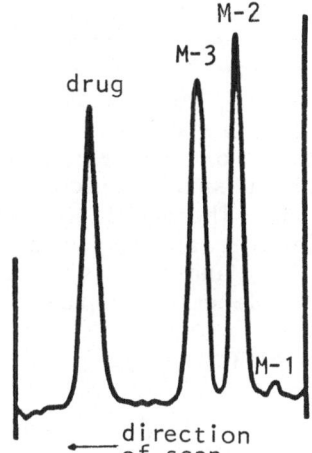

Fig. 1. Metabolites of nafazatrom
in human urine after oral admini-
stration of 10 mg|kg.

taneously, the nafazatrom solutions applied as standards were replaced by spiked plasma or urine samples. The plates no longer had
to be run with an acid-containing solvent, as the degradation of
authentic compound on the HPTLC plate parallelled that of the drug
in test samples and thus obviated imprecision due to this cause. The
Autospotter in conjunction with HPTLC plates was, then, a break-
through in the assay of nafazatrom in body fluids, because accuracy,
reliability and number of samples assayed per day rose markedly.

The sensitivity was 5 ng/ml either for UV densitometry or after
staining. However, the derivatization minimized non-specificity due
to plasma or urine components that interfered in the UV approach.

A recently developed HPLC assay for nafazatrom has a sensitivity
of about 50 ng/ml and thus is inadequate for determining the drug at
low ng/ml levels in plasma.

(4) Imidazole antimycotics

This group of compounds, comprising *N*-substituted imidazoles
with the general formula shown, has been extensively studied during
the past 10 years. The following are now considered:

- **clotrimazole** (BAY b 5097; Canesten[R]),
1-(o-chloro-α,α-diphenylbenzyl)-imidazole:-
- **bifonazole** (BAY h 4502; Mycospor[R]),
1-(α-biphenylbenzyl)-imidazole;
- **BAY f 8751** (oral antimycotic),
2-(2,4-dichlorophenoxy)-1-(1-imidazolyl)-
4,4-dimethyl-3-pentanone.

Clotri-
mazole

At the outset, it was feasible to assay clotrimazole and structurally closely related imidazole antimycotics in body fluids by quantitative TLC after post-chromatographic derivatization by the Iodine-Pauly reaction.- On spraying with iodine and heating ~25 min to remove the excess, cleavage of the C-N bond occurs to yield free imidazole (or iodine-substituted imidazole), which then reacts with diazotized sulphanilic acid (Pauly reagent) to give violet azo-dye spots [5]. The reaction could be applied only to clotrimazole analogues lacking an oxygen at the β-carbon; thus it was applicable to BAY d 3983 and to its active metabolite BAY f 5407, which in fact could be detected on the TLC plate by the same colour reaction.-

BAY d 3983

Active metabolite (BAY f 5407)

But the reaction did not work in the case of BAY c 8208 (oxygen at the β-carbon).-

BAY c 8208

Active metabolite

For determining imidazole antimycotics and their microbiologically active metabolites in body fluids, a more generally applicable derivatization reaction was wanted. GC was at that time insufficiently sensitive (the specific nitrogen detector was then unknown), and HPLC had not (in the early 1970s) become available.

The problem was solved with the discovery of picryl chloride for derivatization, accidentally come across by R. Stegh and later developed to become the most sensitive amongst HPTLC assays with a post-chromatographic colour reaction.

The procedure is as follows. After chromatography the HPTLC plate is sprayed with, or dipped into, a solution of picryl chloride and, after evaporating off the organic solvent, the plate is sprayed with water followed by heating in an oven at 100° for 3-5 min to yield yellow spots of the picrates on a white background. The plate has to be scanned (at 420 nm) within 1 h, because the background - not the spots - is unstable and changes to yellow, thus increasing the background noise and decreasing the sensitivity of the assay.

To summarize, post-chromatographic derivatization of imidazole antibiotics with picryl chloride has the advantage of being applicable generally to N-substituted imidazoles, and thus to all imidazole antimycotics used at present in antimycotic therapy. Moreover, there is applicability to all microbiologically active metabolites, e.g. the main metabolite found after oral administration of BAY f 8751 was identified and assayed expeditiously.-

BAY f 8751:
Cl
Cl—⟨benzene ring⟩—O-CH-C-C(CH$_3$)$_3$ (with C=O, and CH$_2$ attached to imidazole N)

Main metabolite:
Cl
Cl—⟨benzene ring⟩—O-CH-CH-C(CH$_3$)$_3$ (with OH, and CH$_2$ attached to imidazole N)

A further advantage of the approach is the much superior sensitivity compared with GC or HPLC, evidenced by some detection limits:
- clotrimazole, 1 ng/ml (cf. UV densitometry, 50 ng/ml!);
- bifonazole, 0.2 ng/ml (cf. UV densitometry, 10 ng/ml!);
- climbazole, 0.5 ng/ml.
The sensitivity per spot is actually in the pg range, because only 10% of the extract from 1 ml of plasma is spotted. For clotrimazole 100 pg per spot is detectable, for bifonazole 20 pg, and for climbazole 50 pg.

In contrast, GC assay of climbazole has in our hands a sensitivity of only 20 ng/ml (nitrogen detection), and the sensitivities of HPLC assays are, for the following imidazole antibiotics with a similar structure, miconazole 250 ng/ml [6], econazole 40 ng/ml [7], and ketoconazole 100 ng/ml [8].

Another important advantage of the HPTLC assay is that there is immediate recognition of the presence of metabolites, which are generally more polar than the parent drug and have a lower R_f value, whereas in GC or HPLC one merely has an additional peak, which alternatively could be due to endogenous components.

Pharmacokinetic studies on imidazole antibiotics have borne out the power of the reliable HPTLC approach with final derivatization, effective in the low-ng range. Thus, in 6 volunteers following the third of 3 daily vaginal applicateions of 200 mg of clotrimazole as a tablet, serum samples taken at 2, 4, 8 and 24 h mostly showed ≤1ng/ml but some showed several ng/ml; the respective values in the person with the highest levels were 5, 6, 6 and 4 ng/ml. A transdermal absorption study in patients given bifonazole as a 1% cream (applied to ~50 cm^2) showed, throughout day 14, an average plasma level of about 2.1 ng/ml. A third example concerns a tolerance study on a new antimycotic, BAY l 9139, where 0.25 mg/kg was given orally. Even beyond 4 h, after first-order elimination (β-phase; mean plasma $\frac{1}{2}$-life 1.7 h), we were able to discern from 24-32 h values a late (γ) phase of elimination (apparently first-order; mean $\frac{1}{2}$-life 12-15 h), where the actual values assayed were ~5-10 ng/mL. The pharmacokinetic parameters thus found were confirmed in later studies with 5 mg/kg oral dosage.

References

1. Ritter, W. (1977) *Curr. Med. Res. Opin. 4*, 564-573.
2. Ritter, W. (1977) *J. Chromatog. 142*, 431-440.
3. Seuter, F., Busse, W.D., Meng, K., Hoffmeister, F., Möller, E. & Horstmann, H. (1979) *Arzneim.-Forsch. (Drug Res.) 29*, 54-59.
4. Vermylen, J., Chamone, D.A.F. & Verstraete, M. (1979) *Lancet i (no. 8115)* 518-520.
5. Ritter, W., Plempel, M. & Pütter, J. (1974) *Arzneim.-Forsch. (Drug Res.) 24*, 521-525.
6. Sternson, L.A., Patton, T.F. & King, T.B. (1982) *J. Chromatog. 155*, 223-228.
7. Brodie, R.R., Chsseaud, L.F. & Walmsley, L.M. (1978) *J. Chromatog. 155*, 209-213.
8. Alton, K.B. (1980) *J. Chromatog. 221*, 337-344.

Appendix

COMPOSITION OF COLOUR REAGENTS

4-Dimethylaminocinnamaldehyde:- 2.0 g in 100 ml of 6 M HCl, and dilute with 100 ml ethanol; stable when stored in a refrigerator. Spray reagent made by diluting 10 ml with 80 ml ethanol; keeps for some hours at room temperature.

4-Dimethylaminobenzaldehyde:- 2% in ethanol.

Picryl chloride:- 1.5% in toluene; stable for weeks. Dipping solution made by diluting 6 ml with 100 ml of *n*-hexane; prepare daily.

#NC(D)

NOTES and COMMENTS relating to

Determination of particular drugs and metabolites

Comments related to particular contributions:

#NC(D)-1

A Note on

SELECTIVE COLUMN-EXTRACTION OF PSYCHO-ACTIVE
DRUGS AND METABOLITES

R.G. Muusze

Sint Joris Gasthuis (Mental Hospital)
Sint Jorisweg 2
2612 GA Delft, The Netherlands

The determination of psycho-active drugs in blood commonly starts with a liquid-liquid extraction [1-3]. Selective extraction of metabolites can be done with different organic solvents at different pH values [4]. Solvent extraction from serum has two drawbacks: the yield varies considerably (40-80%), and the cleaning and preparation of glassware is time-consuming. Drug concentrates, for HPLC analysis, can alternatively be obtained by disposable RP columns from diluted serum. With Waters SEP-PAK C18 columns (10 × 10 mm), the following procedure is effective.-

1. Wet the column with 5 ml ethanol and 5 ml water.
2. Apply the sample - 1 ml serum + 4 ml pH 10 buffer + i.s.
3. Rinse the column with 10 ml water.
4. Dry the column with nitrogen, 6 ml hexane, and again nitrogen.
5. Elute with 4 ml diethyl ether (drugs) or acetone (drugs + metabolites; but see below), either solvent being readily volatilizable.
6. Evaporate eluate for HPLC; dissolve residue in 100 µl mobile phase.
7. Perform NP-HPLC (Si60) [1]: acetonitrile; 254 nm; 50-100 µl loaded.

The recovery of antidepressants (100 ng in 5 ml water) from disposable columns was checked with different solvents; it was nil with water or hexane or, for nortriptyline, with ethyl acetate (10% for amitriptyline), acetone (20%; similar with dichloromethane) or ether (65%). For nortriptyline the yield with ethanol or butanol was 20% (amitriptyline: 80, 65%). Imipramine yields were somewhat below those for amitriptyline (50% with ether). In other extraction trials with acetone and ether, 'acid' and 'alkaline' modes were used (heading to Table 1), and spiking was done with serum as well as water, giving lower yields possibly because of protein-binding. Data are also tabulated for the HPLC background signal, dependent on the chromatographic system [1] and on 3. and 4. in the disposable-column procedure.

Table 1. Extraction of psychotropic drugs from water (or serum, in parentheses) with disposable RP columns: % yield, and background signal, ng equivalent of compound. 'Acid' and 'alkaline' refer to the serum (taken to pH 2 or 10). Where, with acidified samples, acetone was the eluent, it was pre-acidified (conc. HCl, 0.1% by vol.).

Type	Compound	Extraction yield Acetone		Ether		Bkgd. signal Acetone	Ether
		Alk.	Acid	Alk.	Acid	Alk.	Alk.
Antidep-ressants	Imipramine	75 (45)	90 (60)	60 (50)	15	2 (2)	1 (1)
	Amitriptyline	65 (45)	90 (60)	65 (60)	25	2 (2)	1 (1)
Metabol-ite	Nortriptyline	55 (50)	90 (60)	20 (20)	20	0 (0)	0 (0)
Neuro-leptics	Perazine	90 (75)	35	50	15	0 (2)	–
	Fluphenazine	90 (60)	90	50	–	1 (1)	1 (1)
	Perphenazine	90 (45)	90	50	–	1 (1)	1 (3)
	Thioridazine	60	90	40	15	1 (2)	–
Metabol-ites of thiorid-azine	sulphone	90	80	50	4	1 (1)	–
	ring-SO	80	80	20	0	0 (0)	–
	side-SO	60	60	15	0	0 (0)	–

In the alkaline or acid mode the extraction yield is generally higher with acetone than with ether (Table 1), and the background signal in the alkaline mode is acceptably low. For unknown reasons the background signal from acid-pH samples with acetone extraction is 5-15 ng (not tabulated), which is too high for most psycho-active drugs. The procedure listed on the previous page can be used for drugs whose serum concentration exceeds 10 ng/ml. The background signal is still too high for fluphenazine and perphenazine. Since conditions can be chosen to give a detection limit of 1-2 ng/ml for phenothiazines, liquid-liquid extraction can in general be superseded by column extraction as a fast reliable procedure.

References

1. Muusze, R.G. (1980) in *Phenothiazines and Structurally Related Drugs* (Usdin, E., Eckert, H. & Forrest, I.S., eds.), Elsevier, Amsterdam, pp. 125-128.
2. Miyazaki, K., Arita, T., Oka, I., Koyama, T. & Yamashita, I. (1981) *J. Chromatog. 223*, 449-453.
3. Gillespie, T.J. & Sipes, I.G. (1981) *J. Chromatog. 223*, 95-102.
4. Whelpton, R. & Curry, S.H. (1976) *J. Pharm. Pharmacol. 28*, 869-873.

#NC(D)-2

Analytical case history

A Note on

DETERMINATION OF ATRACURIUM AND ITS METABOLITES

E.A.M. Neill

Department of Drug Metabolism
Wellcome Research Laboratories
Beckenham, Kent BR3 3BS, U.K.

Requirement — *Procedures to assay blood, bile and urine for atracurium – a difficult bis-quaternary compound (stereoisomer mixture) – and products therefrom.*

End-step — *Cation-exchange HPLC (Partisil SCX column, 250 x i.d. 4.6 mm) with fluorimetric detection (280/336 nm excitation/emission). Eluent (3 ml/min): acetonitrile/pH 2 H_2SO_4, 1:1 by vol., in 25 mM $NaNO_3$, 60°.*

Sample preparation — *Plasma (blood spun for 30 sec in Eppendorf centrifuge): 2 ml taken to pH $\simeq$ 6.5 (H_2SO_4) and applied to pre-wetted Sep-Pak C-18 cartridge; after wash with 1 ml pH 1 H_2SO_4, elution with acetonitrile/pH 2 H_2SO_4, 4:1 by vol.*
Bile and urine: equal vol. of acetonitrile added, and filtrate used for HPLC injection.

Comments — *Transformation of the drug by Hofmann elimination (non-enzymic) and ester hydrolysis (enzymic or (non-enzymic) may occur in vitro as well as in vivo; precautions needed in processing. Initial addition of a compound similar in structure to the drug, as a carrier, minimizes adsorption losses. HPLC must not be so efficient as to resolve the drug stereoisomers. Limit detectable: 50 ng/ml plasma.*

Atracurium is a neuromuscular blocking agent with the structure shown overleaf. It was designed to undergo rapid non-enzymic inactivation at physiological pH and temperature by a process of Hofmann elimination [1], which is advantageous because a drug which relies upon metabolism or excretion to terminate its actions may have a prolonged effect in patients with hepatic or renal insufficiency. In this elimination, a quaternary amine becomes tertiary [2] (Fig. 1).

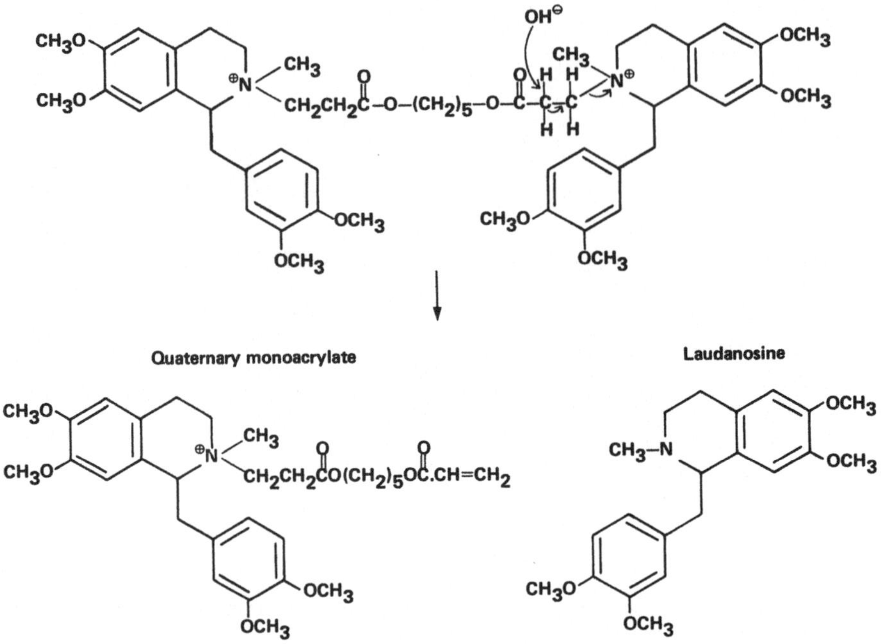

Fig. 1. Atracurium: Hofmann elimination reaction.

Our task was to confirm that the anticipated decomposition products from the above transformations were present *in vitro* and *in vivo*. Assay methods had to be developed for biological samples which would avoid decomposition during processing. Minimal sample preparation followed by HPLC separation was the chosen approach.

SAMPLE PREPARATION

The plasma samples, from various animal species, were frozen if they had to be stored (24 h limit), to minimize the Hofmann elimination reaction. At first acetonitrile was used to precipitate plasma proteins, followed by manual injection of the filtrate. Samples were not very stable.

With the change to C-18 cartridges, sample preparation was rapid, the resulting extract was concentrated instead of diluted, and it was stable in the mobile phase for automatic injection.

Bile and urine were stored frozen after separate collection from anaesthetized cats given a paralyzing dose of radiolabelled atracurium. Marker compounds relevant to identification were run also, spiked into separate portions of bile or urine (e.g. laudanosine).

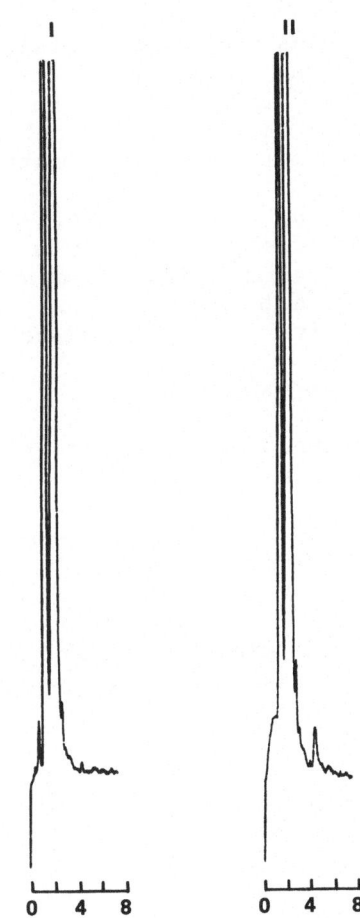

Fig. 2. HPLC of plasma extracts:
I, blank; II, plasma spiked with
atracurium besylate, 50 ng/ml.

Conditions were as stated in the
above précis. Injections 50 µl.
Apparatus: Hitachi 650-LC fluorescence detector in conjunction
with a 1081B HPLC system (Hewlett-
Packard).

HPLC

Fig. 2 shows plasma patterns for the HPLC system finally adopted (see above). Atracurium presents problems in that too efficient
a system gives unwanted resolution of the compound into its isomers,
but efficiency must be adequate to separate it from products derived
from it and from impurities. At first, isocratic conditions using
mixtures of acetonitrile and mineral acids were tried with straight
phase columns. Separation was achieved, except for an impurity that
eluted on the leading edge of the atracurium peak. Column efficiency
was low, only small volumes could be applied, and the mobile phase
was rather corrosive to our HPLC pumps as it contained phosphoric
acid. RP columns were no better. Ion-pairing agents such as benzene sulphonic acid allowed separation of atracurium from all known
impurities; but sample capacity was still low.   Methane sulphonic

acid was used in the mobile phase to separate metabolite peaks under gradient conditions. The later approach based on cation-exchange allowed injection of more sample with a less corrosive mobile phase. Sodium nitrate in the mobile phase (to increase ionic strength) was essential for good chromatography.

The assay has served for clinical studies to ascertain the pharmacokinetic behaviour of atracurium. With the cat samples it was confirmed that Hofmann elimination products appeared as metabolites in bile and urine, as did the possible products from ester hydrolysis (to be published); RP (gradient) with radiolabel was used.

Acknowledgements

Dr. C.R. Jones and Mr. P.M. Johnson are thanked for their help throughout.

References

1. Hughes, R. & Chapple, D.J. (1981) *Br. J. Anaesth.* *53*, 31-44.
2. Geissman, T.A. (1977) *Principles of Organic Chemistry,* 4th edn., Freeman, San Francisco, 1035 pp.

#NC(D)-3

A Note

DISCRIMINATION BETWEEN CARBON MONOXIDE AS A METABOLITE OF CARBON TETRACHLORIDE AND AS A DEGRADATION PRODUCT OF THE BIOLOGICAL MATERIAL

L.J. King

Department of Biochemistry
University of Surrey
Guildford GU2 5XH, U.K.

Carbon monoxide has now been identified as a metabolite of a number of foreign compounds, including tetrahalogenomethanes [1], trihalogenomethanes [2], trichloroethylene [3] and methylenedioxy-phenyl compounds [5]. It is produced as a result of interactions with cytochrome P-450 of the microsomal monooxygenase enzyme system of liver. In several instances, and particularly with CCl_4, the metabolic inactivation by cytochrome P-450 results in loss of haem and initiation of lipid peroxidative damage of the microsomal membranes. Both haem degradation and lipid peroxidation are possible sources of CO, and consequently there is an analytical problem in distinguishing between CO produced as a metabolite of the foreign compound and that which arises from degradation of the biological components.

Both *in vivo* and in incubations *in vitro*, the production of CO can be conveniently followed spectrophotometrically by the formation of carboxy-Hb [1]; but this technique fails to distinguish the source of the CO. This method can also be used to quantify the total production of CO. An alternative method, applicable to incubations, is GC analysis of the headspace gas above the incubation performed in a closed vessel [3]. This method again only determines the total CO.

In our studies on the metabolism of CCl_4 *in vitro* by rat hepatic microsomal fractions under anaerobic conditions [6], the total formation of CO was determined as carboxyhaemoglobin. The formation of CO as a metabolite of CCl_4 was determined using $^{14}CCl_4$. The anaerobic incubations with microsomal fraction, NADPH and $^{14}CCl_4$ were performed in a closed vessel. After incubation, carrier ^{12}CO was added, the reaction mixture was acidified and volatile components were displaced with nitrogen through a cold trap (to retain water vapour), a column of activated silica gel (to retain $^{14}CCl_4$ and $^{14}CHCl_3$) and a catalytic column of 'Hopcalite' to oxidize ^{14}CO to $^{14}CO_2$. The $^{14}CO_2$ was absorbed in phenylethylamine in a tube packed with glass beads, and determined by scintillation counting. Increasing the amount of carrier ^{12}CO 10-fold showed that the ^{14}CO recovery was virtually 100%.

The yield of ^{14}CO over 1 h was 0.8 nmol/nmol P-450. In comparison, the total yield of CO was 1.5 nmol/nmol P-450 for a 10 min incubation, and 2.0 and 3.0 for 20 and 40-60 min respectively. This suggests that under these experimental conditions a major proportion of the CO is produced by degradation of the biological material.

References

1. Wolf, C.R., Mansuy, D., Nastainczyk, W., Deatchmann, G. & Ullrich, V. (1977) *Mol. Pharmacol. 13*, 698-705.
2. Ahmed, A.E., Kubic, V.L. & Anders, M.W. (1977) *Drug Metab. Dispos. 5*, 198-204.
3. Kubic, V.L. & Anders, M.W. (1975) *Drug Metab. Dispos. 3*, 104-112.
4. Traylor, P.S., Nastainczyk, W. & Ullrich, V. (1977) in *Microsomes and Drug Oxidations* (Ullrich, V., et al., eds.), Pergamon, Oxford, pp. 104-112.
5. Yu, L.S., Wilkinson, C.F. & Anders, M.W. (1980) *Biochem. Pharmacol. 29*, 1113-1122.
6. Ahr, H., King, L.J., Nastainczyk, W. & Albrich, V. (1980) *Biochem. Pharmacol. 29*, 2855-2861.

#NC(D)-4   *Analytical case history*

A Note on

HPLC DETERMINATION OF CEFOTAXIME AND ITS METABOLITES

J. Chamberlain, D. Dell* & F.G. Coppin

Hoechst Pharmaceuticals Research Laboratories
Milton Keynes, Bucks. MK7 7AJ, U.K.

The pharmacokinetics of antibiotics are often studied by means of microbiological assays. However this is subject to some uncertainty if the drug is metabolized to species having a different spectrum of antimicrobial activity compared with the parent. For basic pharmacokinetic work therefore it is desirable to use a method which separates parent drug from active metabolites and preferably quantitates the metabolites in addition.

Requirement *Method for determining, in plasma and urine, the new semi-synthetic cephalosporin cefotaxime, its main metabolite desacetylcefotaxime, and minor metabolites (lactones with the β-lactam ring opened).*

End-step *RP-HPLC (C-18) with 254 nm detection, under conditions that are critical if the minor metabolites are to be separated from cefotaxime and other constituents: elution with 10-15% methanol (or acetonitrile) in dil. acetic acid.*

Sample preparation *Deproteinize (plasma) with chloroform/acetone (1:3) and freeze-dry; extract residue with mobile phase.*

Comments *The minor metabolites comprise 10-15% of an oral dose in normal human urine, and are present in plasma from patients with renal insufficiency; but plasma from normal subjects shows only cefotaxime and desacetylcefotaxime in significant amounts. The assay applied to plasma can detect the drug down to 0.5-1 µg/ml. Elution with a higher methanol concentration gives inadequate resolution of metabolites.*

---

* Now at Hofmann-La Roche, Basel, Switzerland

Comments on material in #D

Comments on # **D-1**, J.A.F. de Silva - FLURAZEPAM METABOLITE

Remark by W. Ritter.- A sensitivity of 100 ng/ml hardly seems "good" in view of your earlier work ([7] in #D-1) where you achieved 1 ng/ml by TLC with densitometry. *Reply.-* The context of that work was determination of low levels in plasma, entailing extensive sample preparation including acid hydrolysis to the corresponding benzo-phenone and derivatization for TLC and then fluorodensitometry; the assay was our only means of determining flurazepam and its hydroxy-ethyl and *N*-desalkyl metabolites in plasma following single 30 mg oral doses of the drug. The HPLC-UV approach sufficed for determining the intact benzodiazepine in urine.

Reply to B. Scales.- Only relative bioavailability (oral compa-red with i.v.) as distinct from absolute bioavailability has been looked at. The excretion profile after i.v. administration does not lend itself to direct estimation. Oral dosing leads to extensive first-pass metabolism, affecting the excretion profile although the 'total' metabolite recovery is comparable.

Comments on #D-2, J.B. Houston & J.C. Rhodes - AMINO- & ANTI-PYRINE

Remarks by W. Ritter.- The metabolism of aminopyrine and anti-pyrine isn't really well documented, although they have long been in use; only recently have the hydroxylated metabolites been found, and their fate is uncertain. The chemical instability of the hydroxyla-ted metabolites of antipyrine needs explanation; one wonders if the heterocyclic ring opens up. *Answer to* R.G. Muusze.- The drug role in testing concerns assessing disease status, e.g. liver intoxication.

Comments on #D-3, J. Rosenthaler et al. BROMOCRIPTINE IMMUNOASSAY

Remarks by M.J. Stewart.- Imprecision may arise largely from the charcoal separation step. An alternative approach might be a double-antibody or solid-phase antibody with γ-labelling, e.g. ^{125}I-iodocriptine. *Answer.-* Iodine label is too unstable, but we now have higher s.a. bromocriptine; moreover *(a further answer)* we are able to study in normal as well as in hypoprolactinaemic patients prolac-tin suppression and its relationship to bromocriptine.

Question by A.M.Symons.- If there is indeed a biological test
for bromocriptine, does immunoreactivity run in parallel ? *Reply.-*
I know of no radioreceptor assay whereby bromocriptine could be
assayed.   Since it has dopamine-receptor agonist activity a receptor
assay using, e.g., rat whole-brain homogenate would be feasible.   On
the other hand, measurement of the biological activity of bromocriptine in lowering prolactin secretion would be the more reasonable
approach, and here there does appear to be parallelism, from a study
in man: bromocriptine levels in plasma inversely reflected lowering
of plasma prolactin below lactational levels.

Aspects of the assay: answers to L.E. Martin, J.A.F.de Silva,
J. Vessman.- We agree that 70-80% alcohol is an appropriate diluent
for high s.a. tritium compounds.   The time span from being faced
with setting up an assay to finalizing the RIA was about 3 years.
In considering whether HPLC with fluorescence detection could furnish a sensitive assay (cf. the high fluorescence of the ergot alkaloids), one has to distinguish between the hydrogenated and the non-
hydrogenated ergot peptide alkaloids.   The former have less intense
fluorescence maxima than the latter.   Already, in fact, HPLC, GC and
mass fragmentography were in use to determine bromocriptine in plasma
from patients with Parkinson's disease, and with the high doses used
it is a matter of being able to detect several ng/ml.   Since very
low doses inhibit prolactin secretion, plasma levels are generally
below what can be detected chromatographically.

Comments on #NC(D)-1, R. G. Muusze - EXTRACTING PSYCHOACTIVE DRUGS

Answer to J.A.F. de Silva.- The UV-detection approach is simple
and direct, and suffices for therapeutic monitoring; post-column
derivatization to give a fluorogen would be unnecessarily elaborate.
Remarks by B. Goodwin and U.A. Th. Brinkman.- Non-elution of basic
compounds by hexane in the RP system could, even if [as was the case]
detergent was absent, reflect ion-exchange rather than RP, if protein
in the sample were held by van der Waals forces to the RP material, so
'fixing' the bases. Hexane served merely to remove water from the
column; it is relevant that phenothiazines such as thioridazine are
so insoluble in hexane that it can be used for solvent-segmented
flow systems.   When, in the method used, the compounds were eluted
from the Sep-pak column and dried down, the residue was dissolved in
the HPLC mobile phase and applied *in toto*.

Query by W. Dieterle.- One wonders what is the explanation
(ion-pairing ?) of the effective elution with acetone in the 'alkaline and acid' mode but not with neutral samples and non-acidified
acetone.

Comments on #**NC(D)**-2, E.A.M. Neill – ATRACURIUM AND METABOLITES

Points raised by A.S. Papadopoulos, J. Vessman, L.E. Martin.-
Adsorption sites were effectively swamped by adding as a carrier, in
excess, a compound very similar to the drug in structure but non-
fluorescent under our conditions, such that it had been rejected for
i.s. use.   Silanization of glassware was not tried, but it is known
with some drugs that it can actually cause problems.   Concerning
the action of the drug, the neuromuscular blockade lasts ~15-20 min.
Another example of adsorptive losses, in answer to I.D. Watson.-
With 7-hydroxychlorpromazine (see #E), D. Stevenson & E. Reid found
that the tendency to adsorb onto glass, and maybe 'creep out' in a
subsequent assay with the same tubes, was more effectively overcome
by pre-treatment with chromic acid than by silanization.

Some observations concerning unstable analytes

Examples mentioned (no publication text) by J.W.Gorrod (cf.
[1], and Index entries, e.g. Lability, in this & previous vols.).- C
atoms can be the site for formation (at double bonds) of epoxides,
which may spontaneously give rise to phenols, dihydrodiols, or nuc-
leophil (e.g. GSH) reaction products; or in the case of an α-carbon
adjoining a hetero-atom or an aliphatic amino group there can arise,
by metabolic attack, a carbinol which, respectively, may break down
to give dealkylated compound and aldehyde, or furnishes aldehyde (or
ketone) and ammonia.

Metabolic attack at N atoms can give diverse unstable products,
as shown especially by the Chelsea College group ([1, 2] & #T-5 in
Vol. 5). After *N*-hydroxylation there can arise (maybe favoured by
acidity) ketones (aliphatic parent compounds, via oxime), nitroso
compounds (aromatic), or o-amidophenols (amide, via hydroxamic acid).
The *N*-oxides arising from tertiary amines can become dealkylated, or
acylated to unstable products, or may revert (e.g. on a GC column)
to the parent molecule.   Sulphoxidation may be another 'spontaneous'
process that calls for alertness (see #E).

J.W. Gorrod, *dialogues with* A.S. Papadopoulos, J.A.F. de Silva.-
To protect postulated metabolites it may indeed be worth adding sta-
bilizers, e.g. oxygen scavengers or EDTA.   Where a molecule has more
than one N atom, the most basic is likely to be the site of *N*-oxide
formation; this would accord in the case of perazine-*N*-oxide with
assignment to the ring N bearing a methyl group [3]. Since, however,
the latter resists removal by SO_2 [3] whereas pethidine *N*-oxide (likewise having cyclic N) is thereby demethylated [4], the assignment is
open to question, insofar as steric hindrance could hardly be the explanation.

[Refs. overleaf

1. Gorrod, J.W. (1976) *Proc. Anal. Div. Chem. Soc. 13*, 352-353.
2. Beckett, A.H. (1978)   *Biological Oxidation of Nitrogen* (Gorrod, J.W., ed.), Elsevier/North Holland, Amsterdam, pp. 3-14.
3. Breyer, U. (1969) *Biochem. Pharmacol. 18*, 777-788.
4. Mitchard, M., Kendall, M.J. & Chan, K. (1972) *J. Pharm. Pharmacol. 24*, 915.

Comments on #**NC(D)**-3, L.J. King – CO, AS A METABOLIC PRODUCT

   Remarks by J.S. Oliver.- My own experience with workers exposed to controlled levels of dichloromethane in a factory environment has shown no difference from a normal population in carboxy-Hb levels. (L.J. King was not surprised.) *Parent-molecule assay: ref. supplied by Ed.-* An assay (by GC) for carbon tetrachloride and chloroform in rat blood has been reported [1]; it applies to air also.

1. Reddrop, C.J., Riess, W. & Slater, T.F. (1980) *J. Chromatog. 193*, 71-82.

Comments on #**NC(D)**-4, J. Chamberlain et al.- CEFOTAXIME ASSAY

   Remarks by I.D. Watson.- The poor UV detectability of cephalosporins has led to trial of electrochemical (oxidative) detection, giving a sensitivity of ~1 µg/ml (~100 ng on column). F.G. Coppin, *replying to* M. Brett.- The discovery of unsuspected HPLC•components was a fortuitous outcome of lowering the mobile-phase solvent content.

Metabolites in monitoring routines· discussion initiated by P. Toseland

   Chromatography is vital for disclosing metabolite peaks that may be unexpected, or may be clinically informative as applies in a few instances (e.g. phenobarbitone from dosed primaclone, the epoxide from carbamazepine, and *N*-desmethyldiazepam from diazepam). RP-HPLC could lead to obsessive furnishing of metabolite values in patients. In toxicological screening (M.E.J. Vince) TLC serves for quick rough checking; with tricyclics especially, which may well have been prescribed to some persons being investigated for possible over-dosage, the question of whether the latter really exists and is unrelated to tricyclic presence as evidenced by a TLC spot might arguably be clarified by ascertaining the ratio of parent drug to metabolites. In HPLC (unlike immunoassay methods) specificity for the parent drug is achievable, and (A.F. Fell) will become more assured with the advent of powerful microcomputer-based UV-visible detectors. P. Toseland agreed that this development (e.g. the Hewlett-Packard 8450A) could be relevant to 'defensive toxicology' to confirm conclusions of possibly contentious character.

Senior Editor's plea to the HPLC fraternity.- If an eluent was made by adding 4 vol. of A to 1 vol. of B, don't term it "80% A", taking additivity for granted!

#E

METABOLITE METHODOLOGY: THE STATE-OF-PLAY

E. Reid*

Wolfson Bioanalytical Unit, Robens Institute
of Industrial & Environmental Health & Safety,
University of Surrey, Guildford GU2 5XH, U.K.

Examples of metabolite isolation or determination, partly from this book, are collated and annotated. Useful tools include TLC, GC and, increasingly, HPLC - notably RP-HPLC, though ordinary silica sometimes works better. Ionic additives may, empirically, improve the peak pattern; this may be poor for a conjugate and its parent compound in a single isocratic run. Immunoassay may be influenced by cross-reactivity of metabolites if not pre-separated. Sample preparation is briefly dealt with, as are interferences, losses and artefacts associated with the biological matrix, with solvents, or with the analyte (e.g. 'OH-CPZ' is adsorption-prone). For an especially difficult analyte, 'SCMC', the approach entailed phenylthiohydantoin formation. Conjugate types are listed, and hydrolysis is touched on.

This concluding article serves partly to 'sign-post' and supplement material earlier in this book, and partly to reinforce points of methodology such as have featured earlier in this 'Analysis' subseries (Vols. 5, 7 & 10; listed opposite title page). For retrieval of guidance on particular aspects, the reader may find it advantageous to consult, in this and previous volumes, both the Contents list and the carefully compiled Subject Index.

STRATEGIES FOR ISOLATING OR DETERMINING METABOLITES

A review published elsewhere [1] by Martin & Reid ('Isolation of Drug Metabolites') offers the advantages of general perspective, of comparisons amongst characterization methods, of elementary exposition of sample handling, and of separation examples that complement those given in Table 1. This Table partly serves as a 'peg' onto which some points that follow are hung, e.g. concerning conjugates.

---

* now at 72 The Chase, Guildford GU2 5UL (as Executive Trustee of Guildford Academic Associates – now the Forum/book 'responsible body').

Table 1. Examples of procedures for bioactive molecules where metabolites were separated (*italicized* if a glucuronide – abbreviated *gl'ide* – or other conjugate). The layout is on the same lines as Table 4 and (HPLC) Table 5 in ref. [1], which gives different examples (mainly 'investigative'). The emphasis is on assay procedures, were identities are already known. B denotes blood (usually plasma), and U denotes urine; B_h, U_h signifies human; hyd. denotes hydrolysis.

Parent compound	Material & metabolite(s)	Outline of procedure, & Remarks (if not in text)	Ref.
Neuroactive basic drugs (alkaline pH for solvent extraction)			
Tricyclics, e.g. amitriptyline	B_h: N-desMe (various; e.g. desMe-doxepin)	Simple extrn. (see text), then GC-MS(CI) on KOH/Carbowax (good for desMe's); GC-MS(EI) less effective	2
Phenothiazines, Thioxanthenes	B_h: Various, especially N-desMe; mesoridazine	Et$_2$O extrn.; HPLC (-CN; radial compression columns), maybe electrochem. det.- Pilot study; incl. butaperazine	3
Trazodone*	Brain, B: 1-m-chlorophenyl-[piperazine]	Benzene extrn.(pre-treat if brain); GC-ECD (deriv.)† or -AFID*	4
Various N-containing drugs			
Carbamazepine	B_h: 10,11-epoxy	CH$_3$CN deproteinization; HPLC (C-18) – cf. GC instability	5
Procainamide	B_h: N-acetyl ('active')	CH$_2$Cl$_2$ extrn.; HPLC (C-18) – a 'Selected Method'	6
Ethosuximide etc.	U_h: *gl'ide* of -OH deriv.	Extrn., derivatizn. & GC-MS (usually EI) – a survey	7
l-α-Acetylmethadol	B_h: nor- & dinor-acetyl	1-Cl-butane extrn.; res. derivatized for GC-ECD	8
Cephalothin	U_h: desacetyl	HPLC, direct: anion-exch.; room temp. O.K. Cefotoxitin [similarly.	9
Zimelidine	B_h: N-desMe	Extrn. steps (3); GC-ECD	10
PentaMe-melamine	B_h, U_h: N-desMe (various)	EtAc extrn. on cellulose support; HPLC (C-18)	11
6-Mercaptopurine	Cells: thio-gua nucleotides	HClO$_4$, then KOH; HPLC (anion-exch.) after KMnO$_4$ step	12
Tamoxifen	B_h, tissues, CSF: 4-OH ('active') & N-desMe	Et$_2$O extrn. (tissue homogenate may need protease); HPLC (C-18, ion-pairing) after irradiation (see text)	13
Antrafenine*	B_h,U_h: acid moiety of ester†	Et$_2$O extrn.; HPLC (C-18)*; or GC-ECD (deriv.)†	14

Theophylline	U_h: Various Me-purines	Ion-pair extrn.(+ $AmSO_4$); HPLC (C-18),grad., ion-pairing	15

Drugs lacking a N atom

Compound		Method	
Digoxin	U: mono- & bis-digitoxide incl. dihydro (conjugated)	Extrn. with CH_2Cl_2 containing 3% heptafluorobutanol after hyd.; pentanol concentrate used for HPLC (C-8)	16
Digoxin	Fluids: dihydrodigoxin & other non-polar derivs.	EtAc extrn.; kieselguhr/solvent extrn./TLC steps; NP-HPLC (better, with the simple first extrn., than RP)	17
Guaiphenesin	B_h: β-(2-MeOPhO)lactic ac.	Et_2O extrn.; HPLC (C-8)	18
Clofibrate	B_h,U_h: Clofibric ac. & gl'ide	CH_3CN deprot.; HPLC (C-18)-gl'ide needs separate run	19
Ethynodiol acet.	U_h: diols etc.(conjugated)	Et_2O after hyd.; 'SSEC' (see text); GC-ECD on derivs.	20
Spironolactone	B_h,U_h: canrenone (desAc-thio)	$CHCl_3$ (hyd. if U); HPLC (silica, i.e. NP)	21

Naturally occurring compounds

Steroids (various)	U_h: diols etc., incl. gl'ides	C-18 silica extrn.; derivatizn., Lipidex, capillary GC+prof-ile	22
Oestriol	U_h: gl'ides & sulphate	XAD-2, 20-60% MeOH to elute; HPLC (C-18, ion-pairing)	23
Catecholamines	U_h: 3-O-Me (incl. -tyramine)	Cation-exch. after hyd.; Et_2O extrn.; GC-ECD on derivs.	24
Serotonin	Brain (e.g.): 5-OH-indoles	C-18 silica extrn.; HPLC (C-18; electrochem. detector)	25
Benzyl isothio-cyanate (in plants)	Liver/kidney incubates, or U: mercapturic acid	Deproteinize, then TLC; or (U)EtAc extrn., freeze out water, dry down & ppt. as dicyclohexylamine salt	26

Foreign compounds other than drugs

Aflatoxin B_1	B, if rat: aflatoxicol etc.	$CHCl_3$ (some unextrd.; some protein-bound): TLC, MS, etc.	27
2-Acetylamino-fluorene (AAF)	Liver incubates: various, e.g. OH derivs.	Acidified isopropanol extrn.; HPLC (C-8) with desferal mesylate added, e.g. to reduce N-OH-AAF losses	28
Benzo(a)pyrene	similar to above (AAF)	EtAc extrn. (see text); HPLC (C-18; % MeOH rising from 0%)	29
Parathion	similar: O replacing S, etc.	Acetone deprot.; TLC, autoradiography (for pathways)	30
[Cyclamate →] Cyclohexylamine	U etc.(species diffs.): various, some conjugated	Continuous Et_2O extrn. after hyd.; derivatizn., then GC-FID or GC-MS; paper chromatog. in pilot work (^{14}C)	31

Table 1 also serves to show what can be done, given the skill, in a hospital or toxicological laboratory on a semi-routine basis in res- pect of chemical assays that help, for example, to optimize drug levels in patients. This aspect (outside the scope of [1]; cf. p. 254) is implicit in some of the following articles. Some metabolites hardly matter.

Many articles in the book, and a few Table 1 citations (e.g. [30, 31]), concern delineating metabolic products and hence pathways. For such studies, well indexed elsewhere [32], and for quantitation where identity is known [cf. #A-6, NC(A)-1], TLC has by no means been super- seded by GC or HPLC. However, the ascendancy of HPLC and also the escalating quantity (if not quality) of publications on metabolite determination - largely for bioavailability studies - is evidenced by a recent quarterly issue of *Gas & Liquid Chromatography Abs- tracts*: 10 metabolite-related papers were cited in the GC section, and 27 in the HPLC section. Two were 'hybrids', entailing independent use of GC and of HPLC if parent compound and metabolite(s) were both to be determined. The further example [14] in Table 1 is salutary in relation to the general belief that metabolites especially lend them- selves to the RP-HPLC approach. Ion-pairing (p. 261) may help.

Whilst conditions can usually be found for successful running of a metabolite on an HPLC column (maybe ordinary silica rather than bonded-phase), subject to the molecule posing no detection problem, these conditions may have to be different to achieve elution of the parent compound as a sharp peak rather than a late hump. This was the penalty for success in running a glucuronide ([19], Table 1) or, as amplified below, a chlorpromazine metabolite where an ordinary silica column proved best [33]. The GC approach may likewise call for two different runs (e.g. [4], Table 1), maybe differing in derivatization and detection mode rather than in run conditions. The temptation to invest in GC-MS for quantitation purposes is weaker now that HPLC is recognized to be notably versatile (although not, as yet, readily interfaceable with MS; but see #C-6 in Vol. 7).

Consideration is given later to the HPLC scene in general, and to aspects of sample preparation, this being simpler for HPLC than for GC or for non-chromatographic approaches such as were still in use when, in 1976 [34], sample-preparation strategy was reviewed. No further consideration will be given to GC lore, as given (especi- ally the GC-ECD art - Vol. 7) in earlier volumes; a 1973 sketch [35] is still helpful. Time will tell whether the merits of quartz capillary columns will indeed lead to a GC "renaissance" as anticipated in the context of steroid profiling ([22]; *vs.* RIA), and whether the wel- come trend towards 'splitless' injection will be paralled by confident measurement by ECD in capillary GC. As a 'parting remark' before RIA is considered, journals should ban papers that state conditions (e.g. for HPLC) in 'recipe' form without expounding the rationale. Shrewd refereeing might also reduce the incidence of failure to state vital details, as reflected even in HPLC literature where one laboratory's

methodology may not be reproducible elsewhere. Where a good description *is* given, heed should be paid to a remark made in connection
with traditional extraction/fluorimetry methods for biogenic amines
(cf. examples of newer methods in Table 1): "Directions must be followed explicitly since what appears to be a minor modification may
invalidate the procedures" [35a].

Immunoassay methods in relation to metabolites

RIA and kindred methods are obviously of little help in identifying metabolic products, but are valuable in routine work if care is
taken to ascertain which 'active' or other metabolites cross-react
with the antiserum raised to the parent compound. The examples in
Table 2 give food for thought. Even with plasma rather than urine,

Table 2. Examples of immunoassays where it was checked whether metabolites cross-reacted. Assays were by RIA unless otherwise stated.

Primary analyte	Metabolite(s), esp. Phase I	Whether significant responsiveness seen	Ref.*
Tricyclics, various (Dibenzazepines)	Various	Yes: evidently a general phenomenon – review article	36
Flupentixol	Various	One antiserum (sheep): no Other antisera: yes	37
Fluphenazine	7-OH, 8-OH	Yes (a flupentixol antis'm)	37
Fluphenazine	7-OH	No	37
Perphenazine (Flu-phenazine etc. also react)	*N*-desOHEt Those formed by sulphoxid'n	Yes – but an unimportant cpd. No	38
Δ^9-Tetrahydrocanna-binol	Various	Yes (other cannabinoids too): precede RIA by HPLC	39
β-3,4-Dimethoxyphen-ethylamine (endogen-ous; urine extract)	Various, e.g. 3,4-dimethoxy-phenylacetic	No (but an unidentified trace bioconstituent may react)	40
Chlordiazepoxide (in urine; an oxazepam antiserum)	desMe deriv.	Yes (and to OHEt-flurazepam but not to flurazepam), in EMIT assay	41
Phenytoin (plasma)	Glucuronide of *p*-OH deriv.	Yes, in EMIT assay	42
Nomifensine	*N*-glucuronide	Yes, unlike acid-stable gl'ide	43

* The cited ref. may be merely 'indirect': thus, the first two 'ref.
37' entries are papers cited in [37] (Jørgensen; Wiles & Franklin).

glucuronides of Phase I metabolites can elevate immunoassay values especially in uraemic patients, as with phenytoin ([42], Table 2) and aldosterone [44]; a true RIA value for the parent compound may be obtained by solvent-extracting it at the outset [43, 44]. Instances of cross-reactivity are given earlier in this book (#A-4, D-3). As J.A.F. de Silva points out, judicious hapten synthesis confers specificity (V. 5, p.22). Where antisera react with metabolites that are to be estimated, or with interfering compounds, HPLC separation may be performed, and RIA done on collected fractions, as with THC [39].- Cf. a study on neuropeptides (TSH-releasing hormone, Substance P, vasopressin [45])

Some HPLC trends and practices

An alternative to RIA in post-column detection of THC and its metabolites after HPLC separation is to render the analytes fluorescent by intense UV irradiation – a seemingly ill-defined treatment which does give reproducible results [39].   Pre-column application of this stratagem has enabled tamoxifen and metabolites to be measured in plasma down to 0.1 ng/ml ([13], Table 1) after ether extraction. Where a sensitivity of 25 ng/ml for tamoxifen and its active desmethyl metabolite is sufficient, HPLC-UV conditions developed by D. Stevenson (unpublished work) are effective, with a good UV detector.

Generally there need not be recourse to fluorescence detection if the desired sensitivity is no greater than that obtainable by GC-ECD, provided that the analyte has fair UV absorbance; an adequate load quantity can be achieved by solvent-extracting, concentrating, and injecting a greater volume (say 100 μl) than was traditional in the early days of HPLC. Judicious choice of UV wavelength may minimize blank interferences (as in art. #D-2). Use of low-UV detection (190-220 nm) to improve sensitivity, as with urinary oestrogens and their glucuronides (Table 1), may carry the penalty of interferences [23].   For monitoring anticonvulsants (ethosuximide, primidone, phenobarbital, ethylphenylacetamide, carbamazepine, phenytoin; metabolites of primidone and phenytoin ran separately), the choice of low-UV wavelength fell on 200 rather than 194 nm [46].

Electrochemical detection in its various modes (summarized in Vol. 5, p. 51) with cherished electrodes can overcome sensitivity or interference problems, e.g. in the oestrogen field [23].   Examples are to be found in Table 1 and in #D-1.   Expositions and examples, e.g. from P.T. Kissinger and J.A.F. de Silva, are also given in Vols. 5, 7 & 10 (indexed under 'HPLC').

With the ubiquitous adoption of 5-10 μm packing materials, the adjective 'microparticulate' has become virtually redundant. Unfortunately it is hard to discover from manufacturers' literature whether a particular packing is 'capped' so that residual silanol groups groups are blocked, e.g. by TMS treatment. Simple tests are given by Knox [47] in his inexpensive counterpart to a standard HPLC text [48].

The practitioners' pendulum has not swung completely from unbonded silica towards alkyl-silicas. Variants such as CN-silica
may gain in popularity [cf. 3]. Notably, however, 'straight' silica has
found favour in some metabolite studies (e.g. [33] and Table 1), maybe
because of superior resolution or reproducibility. The too-good-to-
be-true economy measure of solvent re-cycling may be a positive aid
to reproducibility, as this sceptical author has come to accept. Any
equilibration problems [cf. 28] may thereby be minimized.

As is amplified in the Preface, designations such as 'adsorption', 'NP' and 'RP' represent a semantic jungle; likewise the term
'ion pairing' ('paired ion' being a commercially inspired synonym)
for an operational stratagem whose mechanism is murky. Empirically,
there may be benefit from having in the eluent an ion - possibly a
detergent ([47]; cf. [18], Table 1), but not necessarily organic - of
opposite charge to the analyte ionic species; the actual running order
may thereby be altered. In monitoring anticonvulsants, generally
non-ionized, rival stratagems that are claimed to aid resolution or
reproducibility consist of warming the column and of 'ion-pairing'
(with 'TBAP')[46,49]. The usual context nowadays is RP-silica (liquid
stationary phases are obsolete); but there is an increasing tendency
to supplement NP systems, viz. straight silica + water-miscible mobile
phase, with a pairing agent ('counter-ion' - a term not always used
in this sense), as is exemplified in Scheme 1 overleaf, and elsewhere
(citations in [33]). Polar metabolites may then elute in 'pseudo-RP' order.

Usually the counter-ion is deliberately chosen to be undetectable in the eluate, for the obvious reason that analyte absorbance is
swamped if the eluent is strongly UV-absorbing. One idea which turns
out not to be original (see Table 11.1 in [48]; cf. G. Schill, #D-1 in Vol.
7), but which seems to be in abeyance, is to use in conjunction with an
ordinary silica packing a UV-absorbing or fluorescent ionic additive
which is pre-applied and is absent from the non-polar eluent, into
which it should not bleed: thereby an otherwise undetectable analyte
would be detectable as an ion-pair, assuming high affinity. The con-
ferment of detectability by adding a suitable ion-pairing agent - but
to the emerging eluate, with final removal of the excess - has been
proved effective in conjunction with modules for post-column derivatization and extraction (R.W. Frei and colleagues; p. 93 in Vol. 10).

Other ways of manipulating eluent composition so as to improve
the chromatographic pattern include supplementation with a 'suppressing' ion identical in charge to the analyte, or resort to a gradient
([15, 29], Table 1). maybe with no organic solvent at the outset [29].
Ion-exchange HPLC, maybe without warming [9], still has its exponents.

Any interferences, if not due to impurities, may be eliminated
by a suitable sample-preparation procedure as is considered below.
If ion-pair extraction is used where the analyte is strongly polar
(e.g. [15], Table 1; account in #G-3, Vol. 5), observed distribution

behaviour may have no predictive value for achieving 'ion-pair' HPLC separation.   Nor can endogenous interferences be predicted, except that they will be minimized by use of a non-polar solvent such as heptane for extraction. It would be a boon if a heroic analyst were to publish plasma and urine blank patterns for representative extraction solvents; but for HPLC, more than for GC, maybe the choice of representative column conditions for each would get out of hand.

EXTRACTION BY SOLVENTS OR OTHER MEANS

It was because of plasma interferences that back-extraction and re-extraction, unnecessary when GC assay conditions entailing heptane extraction were adapted for HPLC (Scheme 1), had to be retained when a parallel GC procedure was similarly adapted. The rationale for the weaker alkalinity in the latter procedure (Scheme 1) is that the analyte ceases to be ether-extractable if its phenolic group dissociates. This and other subtleties in the extraction procedures are explained in Vol. 5 (#S-2) by R. Whelpton & S.H. Curry for the fluphenazine family of compounds, akin to the CPZ family. The choice of extraction conditions is a recurring theme in this series of books; examples in this volume include #A-2 and #A-4.

Points touched on by E. Reid in Vol. 10 (p. 20) include the nature and role of '(iso)amyl' alcohol (cf. Scheme 1), and the now popular practice of merely deproteinizing plasma with acetonitrile prior to HPLC injection ([49], & Table 1).

PLASMA

Add NaOH to 0.4 M; extract with heptane containing 1.5% amyl alcohol, and dry down in N_2 stream

CPZ, CPZ-SO, & (if any) metabolites such as des-Me-, dides-Me-CPZ (but *not N*-oxides: CPZ-NO reversion → CPZ could inflate CPZ value)

HPLC on silica with 70% methanol (v/v)/1.5% trifluoroacetic acid/10 mM Na heptane sulphonate

PLASMA

Add NH_4OH to 0.5 M; extract with diethyl ether

CRUDE EXTRACT

Back-extract into aq. HCl, and re-extract into ether after K_2HPO_4/NH_4OH addition (swift!), then dry down

7-OH-CPZ, accompanied by CPZ, some CPZ-SO, etc., *assuming no photodecomposition/adsorption*

HPLC as on left, but 2 mM Na heptane sulphonate (OH-CPZ runs too early if 10 mM as needed to elute CPZ-SO readily)

Scheme 1. Assay of chlorpromazine (CPZ) and metabolites and, in parallel, of OH-CPZ [33]. Start based on S.H. Curry's GC approach (#T-4 in Vol. 5).

The sundry observations that follow have been harvested from papers cited above (cf. Table 1; conjugates will be considered later).
- For tricyclics, benzodiazepines and metabolites in plasma, toluene /heptane/isoamyl alcohol (7:2:1) gives clean GC patterns [2].
- For trazodone in brain as distinct from plasma, benzene extraction is preceded by preparation of a deproteinized homogenate (acetone/M formic acid, 85:15) and heptane/chloroform treatment [4].
- Use of 1-chlorobutane ([8]; undiscussed) may be costly!
- Extraction with the aid of vortexing (into dichloromethane, after NaOH addition, in the example [6]) may lead to a gel, obviated by vortexing in brief pulses and remedied by agitating with a wooden stick. Similar trouble was encountered [33] when, in an attempt to use a single plasma sample for the parallel assays in Scheme 1, the pH of the once-extracted aliquot was lowered and the ether treatment performed.
- To extract tamoxifen efficiently from tissues needed protease treatment [13], as in forensic methodology described in #C-1, Vol. 10.
- Salt in high concentration helps where solvent miscibility with water is appreciable [15; cf. 7]; likewise if the volume of organic phase is kept small so as to concentrate the analyte. With acetonitrile (Vol. 10, p. 27) or 2-propanol, both completely miscible, salt may give 2 phases.
- The advantage of heptafluorobutanol rather than butanol as a solvent constituent [16] lies in its ease of volatilization; this is likewise the case for trifluoroacetic acid which may be an eluent component [33], a deproteinizing agent, or a by-product in derivatizing for GC-ECD where lingering traces are a peril (p.67 in Vol. 7).
- To obtain ethynyl steroids free of non-ethynyl steroids, silver sulphoethylcellulose (SSEC) is an effective agent [20].
- Recovery from incubates after freezing (cf. Vol. 10, p. 25), or by detergent or methanol, was poor for PAHs; but ethyl acetate worked well with unfrozen material [29].

An effective solid agent for extracting acidic and neutral drug metabolites from urine is DEAE-Sephadex [7]. Both for solid agents and for solvents, index entries (particularly 'Adsorbents' and 'Solvent') in earlier volumes may be consulted; in Vol. 10, attention is drawn to p. 23 and to #D-4. There is general guidance in [1,34] and, albeit brief, in [35]. Article #D-7 in Vol. 7 surveys solid agents of help in preparing samples for HPLC. In the present book, relevant articles include #A-2, #A-4 and #C-1. Alumina favours 'vicinal hydroxyls'.

MINIMIZING LOSSES AND OTHER TROUBLES

Some solvents and analytes present difficulties as described in previous volumes, notably in Vol. 10 [E. Reid, #0; J.A.F. de Silva, #D-4; B. Scales, #NC(F)-4]. Besides the following examples, there are relevant observations in preceding articles, e.g. #NC(D)-2, and (in previous vols. also) under the Index entries 'Plasticizers', 'Lability', and 'Storage'. Tricyclics may suffer with some sampling conditions [2].

Chloroform used as bought may not behave identically with chloroform freed from stabilizer, as found in the late-1940s when silica-gel chromatographic runs on derivatized amino acids with chloroform for elution ran into trouble. It may be desirable to supplement washed chloroform with ethanol (say 2%). Traces of phosgene can attack the analyte, directly as when artefactual carbamates arose from tricyclics [50], or indirectly by attacking ethanol in the first instance (citation in Vol. 5, p. 68). Chloroform and also dichloromethane or 1,2-dichloroethane (details of purity regrettably lacking) attacked imipramine, a tertiary amine, to give a quaternary amine ([51]; cf. C.R. Jones, p. 144 in Vol. 5). Dichloromethane may be troublesome if it contains cyanogen chloride (p. 25 in Vol. 10) or, in HPLC, if it is dried down (p. 371 in Vol. 10).

Problems due to analyte instability are exemplified by the tendency of N-oxides in the CPZ family to revert to the parent amine, perhaps merely on exposure to nitrogen (#B-4 in Vol. 7; cf. #T-5 in Vol. 5). Promezathine-5-sulphoxide may revert to promezathine on a GC column [52]. OH-CPZ is especially susceptible to acidity, light, and adsorptive losses (Scheme 1); with unscratched chromic-cleaned glassware, adsorption is minimized.

Adsorption troubles likewise had to be overcome in studies with 'OH-AAF' (Table 1, [28]), norantipyrine (#D-2) and atracurium [#NC(D)-2]. Significant volatility of an analyte can also cause losses, when an extract is dried down (see 'Evaporation' in Vol. 10 Index). It may help to perform the drying in the presence of a trace of acetic acid (for a basic analyte) or of piperidine (for a carboxylic acid) [7]. When a GC method [53] for the neutral drug diethylallylacetamide (and its metabolites) was set up in the author's laboratory (with J.P. Leppard) for the parent drug, this tended to disappear until warming was banned in the step where chloroform was evaporated under a nitrogen stream. (In this study, F.E. Wilson found ethinamate to be a good i.s.; its volatility is similar.)

Problems in developing a method for the sulphide 'SCMC'

Given the common situation that assay literature was lacking for a long-established drug, TLC/ninhydrin offered some hope for the aliphatic compound in question, which was prone to streak, but was abandoned in favour of the novel approach indicated below (legend to Scheme 2). Although the compound is not a metabolite, the pitfalls encountered may be illuminating. The idea was to use, for an amino acid, a derivatization mode which is a tool in protein sequencing, where the N-terminal amino acid is removed as a 'PTH' derivative.

Scheme 2 (opposite). Assay procedure developed by E. Reid [54] for S-carboxymethylcysteine (SCMC), converted into a UV-absorbing phenylthiohydantoin (PTH derivative) - a novel approach entailing difficulties.

COMMENTS

#PLASMA containing endogenous amino
acids and, at comparable level (~10μg/ml),
SCMC: $H_2N-CHR-CO_2H$ where R =
$HO_2C-CH_2-S-CH_2-$

| *Deproteinize at $2°$ with perchloric acid (PCA), and centrifuge*

Storage problem.– SCMC largely lost after few weeks at $-20°$, though std. in weak acid O.K.

#ACID–SOLUBLE FRACTION

| *Aliquot partially neutralized with KOH at $2°$; let $KClO_4$ settle; supernatant (still pH<3) at ~$20°$ →pH>8 with $KHCO_3$, then re-chill*

Deproteinization inefficient if acetone [55] rather than PCA. C–18 silica or cation-exchanger as alternative? – Poor elution of SCMC.

#PERCHLORATE–FREE SUPERNATANT

| *CONDENSATION in a trimethylamine–buffered aqueous acetone medium with phenylisothiocyanate (overnight, ~$20°$; shake well!) to give a phenylthiocarbamate:*

$\underrightarrow{\phi NCS}$ $\phi NH.\underset{NH-CHR.CO_2H}{\overset{}{C}}=S$

then remove ϕNCS: 2 toluene extrns.

Co-precipitation losses can occur, →protein ppt. or else (like Me-dopa [58]) →$KClO_4$ ppt., or with cystine at pH say ~5, $2°$.

Risk to condensation if residual ClO_4^- ions, so chill well!

Plasma blanks 'dirty' if 1 h at $40°$ [55] or if conventional pyridine medium or if acetone >25% v/v [cf. 55]; ϕNCS immiscibility demands vigorous shaking, otherwise **condensation fails.**

#AQUEOUS LAYER

| *CYCLIZATION after sulphuric acid addition (→pH <2) and N_2 flushing: 2 h, $45°$, closed vessel; first product rearranges:*

$\phi NH-\underset{\underset{HCR}{|}}{\overset{}{C}}\underset{\underset{}{CO}}{\overset{}{---}}\overset{}{S}$ $\underrightarrow{-H^+}$ $\underset{\underset{HCR}{HN}\underset{}{CO}}{S=\overset{}{C}---N\phi}$

then optionally add PTHasp as a final-stage i.s.; extract with benzene/EtAc (1:1 by vol.), 15 min

Toluene O.K. in place of benzene [55]; must be twice in either case.

Cyclization yield impaired if no flush-out or if weaker acidity, or if done at ~$20°$ overnight (4 h O.K.); can be $80°$ for pure cpd. (as for phe or tyr in plasma [55]), but with **plasma →'dirty' blank**, nullifying any gain in yield (which is 30-60%).

#ORGANIC LAYER containing PTH–SCMC

| *Dry down: N_2, gentle heat (residue can be kept at $-20°$, N_2); $\frac{1}{2}$-h before HPLC injection, vortex in acidified eluent, $2°$; then dilute with eluent*

Not feasible to introduce at start, as i.s., an amino acid, e.g. asp or S-carboxyethylcysteine: suspected cross-condensation! **Don't replace benzene by toluene:** high drying-down temp.

#HPLC LOAD SAMPLE

C–18 column; 12% v/v CH_3CN in pH 4 acetate buffer; 265 nm. Blank peak pattern has 'window' where PTH–SCMC appears; then late 'junk'.

Stability problems with PTH–amino acids, esp. PTH–SCMC (UV absorbance loss): severe at pH >7; acid pH maybe O.K. for few h at $2°$ unless concn. very low. – Slick HPLC! O_2 not the culprit.

Reproducibility good for HPLC and for whole assay as described.

There may be relevance to assay of cysteine conjugates (cf.#A-1; struct-ural resemblance to SCMC). One pitfall concerned the condensation step.

The starting point was a bald description [55] of the isolation (not for assay purposes) with final HPLC, as PTH derivatives, of two amino acids from plasma. However, unlike pure compounds that abound in PTH literature, plasma proved to be a rich source of interferences. These seemed not to be due solely to endogenous amino acids, which are present at levels comparable to those of the drug and which like-wise give PTH derivatives. The description [55] had to be departed from in respects such as cyclization temperature but was vindicated regarding use of acetone (now 25% only) in the condensation medium; later the precedent (J. Sjoquist; see [56]) was tracked down. Pyridine is a menace!

These and other difficulties are set down in Scheme 2*, along-side the procedural steps finally adopted. SCMC kept quite well in dilute acid, but not at the neutral pH of plasma. This instability, presumably associated with the sulphide group, is akin to the oxygen sensitivity (but acid-stability) of a sulphide isolated from a modi-fied wool peptide [57]. Acetone deproteinization, even with acidi-fication, led to an HPLC 'junk pattern'. Conventional conditions for $HClO_4$ treatment and ClO_4^- removal had to be modified to obviate erratic losses. At the cyclization stage in particular, oxygen ex-clusion was vital. A notable surprise was the instability of dilute solutions of PTH derivatives, especially PTH-SCMC, even at acid pH[†]. Yet there was a successful outcome [54] to this multi-variable study.

CONCLUDING COMMENTS, ESPECIALLY ON CONJUGATES

This survey has touched on various advances and difficulties in the determination of bioactive molecules in biological specimens. Whilst assays on urine remain informative, there is growing interest in plasma levels of drugs and metabolites, especially 'active', e.g. in connection with pharmacokinetics or therapeutic monitoring. As a cautious generalization (cf. Table 1), the aim of assaying parent compound and Phase I metabolites simultaneously may be more readily achievable with N-desalkyl or -desacyl metabolites than with hydroxy-lated metabolites. Immunoassay is a blunt tool in the present context.

Where the administered compound is an ester, its assay conjoin-tly with a carboxylic acid derived therefrom may be difficult (cf. [14], Table 1). However, it is a normal expectation with ester-type drugs that there will be rapid hydrolysis in plasma to give an active acid moiety, which is the species that particularly calls for assay procedures (e.g. [19]; p. 320 in Vol. 7; p. 218 in Vol. 10). It is for this reason that esters are classified along with acids in the Index of Compound Types which complements the General Index.

---

* α-Methyldopa [58] exemplifies losses onto $KClO_4$ (also use of alumina).
† Sensitivity to light or adventitious oxidants[59] could not be blamed.

There is increasing interest in determining the levels of Phase II metabolites (conjugates) besides the unconjugated parent compound, and increasing awareness of diversity in type and behaviour (Table 3).

Table 3. Conjugate types, and index to some examples in the present book. (The Compound-type Index is more systematic in listing any conjugate studied – denoted ^{C} – for particular compounds, but does not indicate the type.) For glucuronidation especially, subtleties concerning types of linkage other than ether (Table 4; arts. #C-1 & #C-2 are informative) are evident only on consulting the cited p. nos. *Adapted, with permission, from Table 2 in ref. [1].*
Ar = aryl; Al = aliphatic or alicyclic; X = electron-attracting, e.g. CO, NO_2.

Type of conjugation	Parent-compound group	Page citations for type
Glucuronidation	OH (commonest), COOH, NH_2, >NH, SH, (rare) C	6–10, 27, 31, 104, 122–129, 131, 132, 150, 156, 161, 181, 184, 191, 196, 204, 208, 210, 268
Sulphation	Ar–OH, Ar–NH_2, Al–OH	4–7, 129–130, 154–157, 210 (enol type: 155)
Methylation	Ar–OH, NH_2, >NH, =N, SH	
Acylation, e.g. acetyl	Ar–NH_2, some Al–NH_2, hydrazides, Ar–SO_2NH_2	6, 182–185
Fatty acid acylation	Ar–OH, Al–OH	
GSH conjugation and mercapturic acid syn.	Ar–X, Al–X, Ar–CH=CH–X, Al–CH=CH–X, Ar/Al–epoxides	4–9, 152, 156, 157
Phosphation	Ar–NH_2, Ar–OH	
Glycine/taurine/glutamine peptide conj'n	Ar–COOH	4, 124, 156, 157, 161–166
Glycosidation/conj'n with other sugars	R–COOH, Ar–OH	152, 163, 164

Along with sulphates, ether-type glucuronides dominate the scene as 'classical' conjugates. Strictly speaking, a 'glucuronide' is by definition a 1-O-linked glucopyranosiduronate (as J. Caldwell has remarked to the author).   The variations that have become apparent (Table 4), e.g. furanose structure, understandably remain under the 'glucuronide' umbrella where mentioned in preceding contributions to this book. Ether-type glucuronides are the context for the well-documented axiom that, for a given parent compound, the glucuronide: sulphate ratio depends on the species. What matters to the analyst is that a polar acidic metabolite, whatever its nature, may be hard to extract even under favourable conditions, e.g. shaking acidified

Table 4. Susceptibility of different glucuronide types to hydrolysis by alkali, acid, or a 'typical' β-glucuronidase (e.g. snail; species differences are mentioned in #A-1). Information largely derived from #C-2 and [1], and may not apply to all aglycones, nor (enz.) to furanose. Attack denoted thus: strong, ++; slow, +; none, 0. 'Alkali' = pH 9 or 10. *Drs. J. Caldwell & L. E. Martin kindly advised; blameless for any errors.*

Type	Alkali	Acid	β-Glucuronidase	Remarks
Ether (Enol similar)	0	+	++, esp. if aromatic	Possible complications, e.g. alk. rearrangements / ester isomers
Ester	++	++	+	
N- (some artefactual)	0	some ++	some ++	Too few examples for safe generalization
S-	0	0	++	
C-	0	?++	some ++	

urine with ethyl acetate [35]. Perhaps a high-salt [17] or ion-pair system may help; but recourse is commonly had to a solid agent [1, 7] although analyte recovery from the agent may not be facile.

One example [19] in Table 1 concerns the isolation of an intact glucuronide without pre-extraction; it shows the usefulness of RP-HPLC in obtaining conjugates as early-running sharp peaks, although under isocratic conditions there may be unduly late elution of unconjugated parent molecules. Sulphates have yet to become good 'HPLC citizens'.

With nomifensine (studied in J. Chamberlain's laboratory; cf. Vol. 10, p. 371) there are two glucuronides which outweigh the active parent molecule in plasma; one (the N-glucuronide) is labile to acid or warmth, although the question of its artefactual reversion to nomifensine needs clearer evidence than one group has presented [43]. Conjugates may, then, be treacherous to the analyst, and the converse situation of resistance to deliberate hydrolysis is now well documented [1]. Susceptibility to hydrolysis (Table 4) may in fact give evidence concerning the nature of an unknown conjugate. Resistant conjugates may yield to unorthodox conditions for hydrolysis, e.g. (cited in [1]) 'solvolysis'. The present revival of interest in conjugates promises clearer guidelines within a few years.

References

1. Martin, L.E. & Reid, E. (1981) *Progr. Drug Metab.* 6, 197-248. [Bridges, J.W. & Chasseaud, L.F., eds.; Wiley, Chichester].
2. Chinn, D.M., Jennison, T.A., Crouch, D.J., Peat, M.A. & Thatcher, G.W. (1980) *Clin. Chem.* 26, 1201-1204.
3. Curry, S.H. & Brown, E.A. (1981) *IRCS Med. Sci.* 9, 170-171.

4. Caccia, S., Ballabio, M., Fanelli, R., Guiso, G. & Zanini, M.G. (1981) *J. Chromatog. 210*, 311–318.

5. Astier, A., Maury, M. & Barbozet, J. (1979) *J. Chromatog. 164*, 235–240.

6. Sawchuk, R.J. & Cartier, L.L. (1980) *Clin. Chem. 26*, 835–839.

7. Horning, M.G., Gregory, P., Nowlin, J., Stafford, K., Lertratan- angkoon, K., Butler, C., Stillwell, W.G. & Hill, R.M. (1974) *Clin. Chem. 20*, 282–287.

8. Henderson, G.L., Weinberg, J.A., Hargreaves, W.A., Lau, D.H.M., Tyler, J. & Baker, B. (1977) *J. Anal. Toxicol. 1*, 1–9.

9. Buhs, R.P., et al. (1974) *J. Chromatog. 99*, 609–618.

10. Larsen, N.E. & Marinelli, K. (1978) *J.Chromatog. 156*, 335–339.

11. Benvenuto, J.A., Stewart, D.J., Benjamin, R.S., Smith, R.G. & Loo, T.L. (1981) *J. Chromatog. 222*, 518–522.

12. Tidd, D.M. & Dedhar, S. (1978) *J.Chromatog. 145*, 237–246.

13. Sternson, L.A., Meltzer, N. & Shih, F.M.H. (1981) *Anal. Lett. 14B*, 583–600.

14. Guinebault, P.R., Broquaire, M., Sanjuan, M., Rovei, V. & Braithwaite, R.A. (1981) *J. Chromatog. 223*, 103–110.

15. Muir, K.T., Jonkman, J.H.G., Tang, D.-S., Kunitani, M. & Riegelman, S. (1980) *J. Chromatog. 221*, 85–95.

16. Eriksson, B.-M., Tekenbergs, L., Magnusson, J.-O. & Molin, L. (1981) *J. Chromatog. 223*, 401–408.

17. Eichhorst, O. & Hinderling, P.H. (1981) *J. Chromatog. 224*, 67–93.

18. Ketelaars, H.J.C. & Peters, J.G.P. (1981) *J. Chromatog. 224*, 144–148.

19. Veenendaal, J.R. & Meffin, P.J. (1981) *J. Chromatog. 223*, 147–154.

20. Lewis, C.J. & Vose, C.W. (1981) *Anal.Proc. 18*, 253–255.

21. Neurath, G.B. & Ambrosius, D. (1979) *J.Chromatog. 163*, 230–235.

22. Shackleton, C.H.L. (1981) *Clin. Chem. 27*, 509–511.

23. Dixon, P.F., Lukha, P. & Scott, N.R. (1979) *Proc. Anal. Div. Chem. Soc. [now Anal. Proc.] 17*, 302–305.

24. Lax, P.M. King, G.S., Pettit, B.R. & Sandler, M. (1979) *Clin. Chim. Acta 96*, 269–272.

25. Falkowski, A.J. & Wei, R. (1981) *Anal. Biochem. 115*, 311–317.

26. Brüsewitz, G., Cameron, B.D., Chasseaud, L.F., Görler, K., Hawkins, D.R., Koch, H. & Mennicke, W.H. (1977) *Biochem. J. 162*, 99–107.

27. Wong, Z.A. & Hsieh, D.P.H. (1978) *Science 200*, 325–327.

28. Smith, C.L. & Thorgeirsson, S.S. (1981) *Anal. Biochem. 113*, 62–67.

29. Krahn, M.M., Schnell, J.V. Uyeda, M.Y. & MacLeod, W.D. (1981) *Anal. Biochem. 113*, 27–33.

30. Neal, R.A. (1967) *Biochem. J. 103*, 183–186.

31. Renwick, A.G. & Williams, R.T. (1972) *Biochem. J. 129*, 857–867.

32. Hirtz, J., ed.(1974) *Fate of Drugs in the Organism*, Vol. 1, Dekker, New York, 393 pp.; also subsequent volumes.

33. Stevenson, D. & Reid, E. (1981) *Anal.Lett. 14B*, 1785–1805 *[not 741–761 – a mutilated version that needs **deletion + cross-ref.** in Library copies, please; the topic is CPZ & metabolites]*.

34. Reid, E. (1976) *Analyst 101*, 1–18.

35. Riedmann, M. (1973) *Xenobiotica 3*, 411–434.

36. Scoggins, B.A., Maguire, K.P. & Burrows, G.D. (1980) *Clin. Chem. 26*, 5–17.

37. Midha, K.K., Cooper, J.K. & Hubbard, J.W. (1980) *Comm. Psychopharmacol. 4*, 107–114.

37a. Anton, A.A. & Sayre, D.F. (1972) in *The Thyroid and Biogenic Amines* (Rall, J.E. & Kopin, I.J., eds.), North Holland, Amsterdam, pp. 398–436.

38. Midha, K.K., Mackonka, C., Cooper, J.K., Hubbard, J.W. & Yeung, P.K.F. (1981) *Br. J. Clin. Pharmacol. 1*, 85–88.

39. Moffat, A.C. (1978) *Proc. Anal. Div. Chem. Soc. 15*, 237–239.

40. Riceberg, L.J. & Van Vunakis, H. (1975) *Biochem. Pharmacol. 24*, 259–265.

41. Poklis, A. (1981) *J. Anal. Toxicol. 5*, 174–176.

42. Aldwin, L. & Kabakoff, D.S. (1981) *Clin. Chem. 27*, 770–771.

43. McIntyre, I.M., Norman, T.R. & Burrows, G.D. (1981) *Clin. Chem. 27*, 203–204.

44. Jones, J.C., Carter, G.D. & Macgregor, G.A. (1981) *Ann. Clin. Biochem. 18*, 54–59.

45. Spindel, E., Pettibone, D., Fisher, L., Fernstrom, J. & Wurtman, R. (1981) *J. Chromatog. 222*, 381–387.

46. Christofides, J.A. & Fry, D.E. (1980) *Clin. Chem. 26*, 499–501.

47. Knox, J.H., ed. (1978 [year unlisted]) *High-Performance Liquid Chromatography*, Edinburgh University Press, Edinburgh, 205 pp.

48. Snyder, L.R. & Kirkland, J.J. (1979) *Introduction to Modern Liquid Chromatography*, 2nd edn., Wiley, New York.

49. Stafford, B.E., Kabra, P.M. & Marton, L.J. (1980) *Clin. Chem. 26*, 1366.

50. Wester, R., Noonan, P., Markos, C., Bible, R., Aksamit, W. & Hribar, O. (1981) *J. Chromatog. 209*, 463–466.

51. Hansen, S.H. & Nordholm, L. (1981) *J. Chromatog. 204*, 97–101.

52. Patel, R.B. & Welling, P.G. (1981) *Clin. Chem. 27*, 1780–1781.

53. Uehleke, H. & Brinkschulte-Freitas, M. (1978) *Arch. Pharmacol. 302*, 11–18.

54. Reid, E., McDonald, T. & Burton, J.S. (1981) *Anal. Lett. 14B*, 615–627.

55. Trefz, F.K., Byrd, D.J., Blaskovics, M.E., Kochen, W. & Lutz, P. (1976) *Clin. Chim. Acta 73*, 431–438.

56. Bailey, J.L. (1967) *Techniques in Protein Chemistry*, 2nd edn., Elsevier, Amsterdam, 406 pp.

57. Friedman, M. (1973) *The Chemistry and Biochemistry of the Sulphydryl Group in Amino acids*, Pergamon, Oxford, 485 pp.

58. Kim, B.K. & Koda, R.J. (1977) *J. Pharm. Sci. 66*, 1632–1634.

59. Blackburn, S. (1970) *Protein Sequence Determination*, Dekker, New York.

Cumulative Index of Compound Types
(covering Vols. 5, 7 & 10 also)

Generic names as in the 'Merck Index' (which gives formulae) are used where applicable for parent compounds such as pre-dominate in the Index. **Superscripts** *signify that metabolite(s) information is to be found: Phase I,* [1] *, or* [1] *if including N-de(s)alkyl derivs.; Phase II (conjugate),* [c] *. Some parent compounds fall under a class title, e.g. Phenylethylamine. In general the 10-category CHEMICAL CLASSIFICATION (overleaf) serves to collate according to analytically relevant common features.* **Style guide:**

- ASSAY GUIDANCE OR CITATION FOR 'REAL' SAMPLES as specified parenthetically (see below): e.g. 25 (bld), 91-(urn) *or, if in a previous vol., v.5: 91-(urn) where '-' denotes a major entry.*

- Merely POINTS RELEVANT TO SAMPLE HANDLING, e.g. extractability: e.g. 17 *or, if in a previous vol., v.7: 17.*

- Merely BEHAVIOUR RELEVANT TO MEASUREMENT (e.g. HPLC t_R *or derivatization): e.g.* 17 *; **not** re-indexed here if in a previous vol. - see Cumulative Index in Vol. 7, and Compound-type Index in Vol. 10.*

SAMPLE TYPE is shown parenthetically (), abbreviated thus:

aq = *water, effluents, beverages*	slv = *saliva*
bld = *blood, usually plasma*	tss = *tissues (animal), incl.*
sld = *solids other than animal*	*incubates/cells/perfusates*
tissues: e.g. faeces, crops	urn = *urine*

CATEGORY I (no amino group, nor cyclic N except maybe imino)

#Ia (acid or ester, *not conjugate*)
  See also Acids *General Index entry & Vol. 7 Compound Index*

Acrylate: 153[c](urn)
Adipic acid: *v.10: 158-(tss)*
Arylacetic acids: 162-[1],[c] (urn, etc.)
Aspirin: 156[c](bld, sld, urn); *v,5: 35 (bld, urn)*

Barnon: *v.10: 141-(sld)*
Benzoic acid: 52 (urn), 156[c](tss)
Bile acids: 156[c] (urn)

#Ia, continued

Buniodyl: 165[c]

Carbaryl: 156[c] (tss, urn)
Carbofuran: *v.10: 155 (sld)*
Carboxylic acids (*not* conjugates), various (see *also* Aryl...., Phenoxy): 50 (urn); *v.5: 123; v.7: 301 (bld); v.10: 75 (aq)*
- phenolic: 47- (urn), 54 (urn), 93; *v.7:77(aq)*
Cephalosporins: see #IIa (even if of #Ia type)
Cephradine: *v.5: 227- (bld), 341- (bld, urn)*

ASSIGNMENT CATECHISM

Note on the term acid: *this signifies pK <6, hence phenols excluded.
Omitted: intact conjugates (superscript ^c after
parent-compound name) as tabulated on p. 267.
Ester-type parent compounds here rank as acids.*

Policy for Phase I metabolites: *only parent compound indexed, with
superscript [1], or [1] if incl. N-de(s)alkyl derivs.*

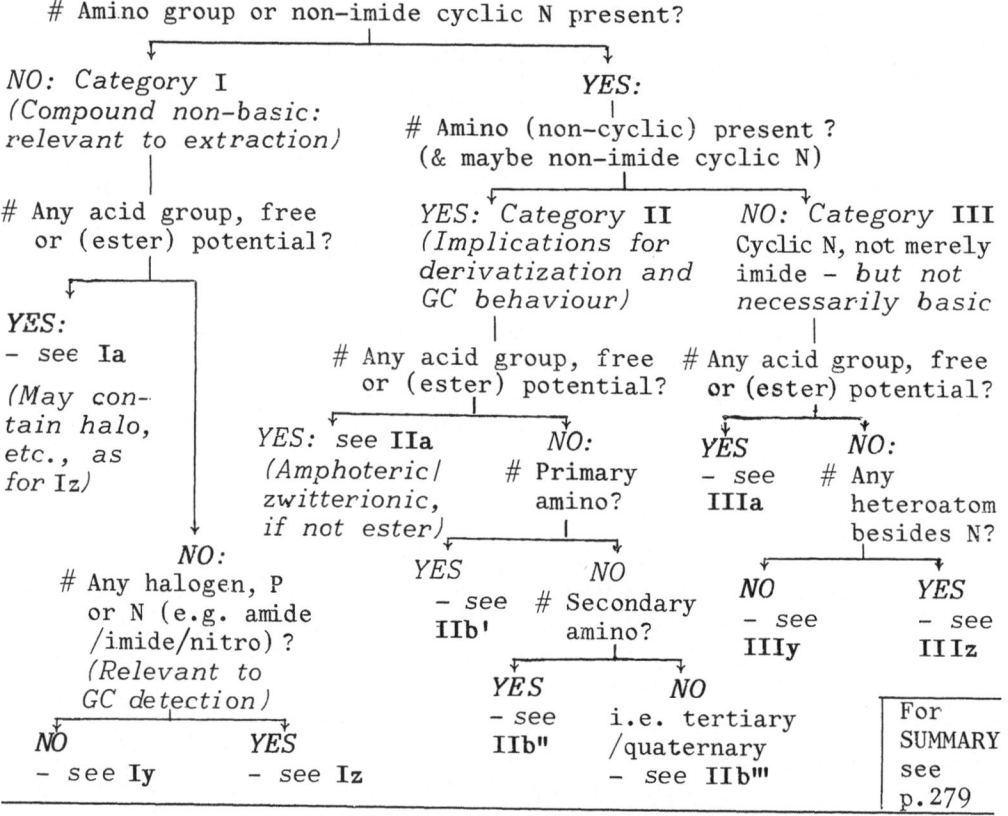

\# Amino group or non-imide cyclic N present?

NO: Category I
*(Compound non-basic:
relevant to extraction)*

\# Any acid group, free
or (ester) potential?

YES:
– see **Ia**

*(May con-
tain halo,
etc., as
for* Iz*)*

NO:
\# Any halogen, P
or N (e.g. amide
/imide/nitro)?
*(Relevant to
GC detection)*

NO – see **Iy**   *YES* – see **Iz**

YES:

\# Amino (non-cyclic) present?
(& maybe non-imide cyclic N)

YES: Category II
*(Implications for
derivatization and
GC behaviour)*

\# Any acid group, free
or (ester) potential?

YES: see **IIa**
*(Amphoteric/
zwitterionic,
if not ester)*

NO:
\# Primary
amino?

YES – see **IIb'**   *NO* \# Secondary amino?

YES – see **IIb"**   *NO* i.e. tertiary /quaternary – see **IIb'''**

NO: Category III
Cyclic N, not merely
imide – *but not
necessarily basic*

\# Any acid group, free
or (ester) potential?

YES – see **IIIa**   *NO:* \# Any heteroatom besides N?

NO – see **IIIy**   *YES* – see **IIIz**

For
SUMMARY
see
p. 279

#Ia, continued

Chlorfenvinphos: 121–[1]c (urn)
Chlorpropamide: see #Iz (though
  pK$_a$<5)
Clofibrate: 6–c (bld, urn), 164–c
  (urn), 257c (bld, urn); *v.7:*
  320 (bld)
Clozic: 191–c (urn), 197c
Cromoglycate: *v.7: 333–(urn)*

#Ia, continued

Crotonate: 153c (urn)
Cypermethrin: 124[1] (urn); *v.10:*
  128[1], 132–(sld)

Dicapthon: *v.10: 99 (aq)*
Dicrotophos (Bidrin): *v.10: 136
  (sld)*

#Iy, continued

Estrogens: see Oestrogens
Ethanol: *v.7: 36 (bld); v.10: 233-
(aq, etc.)*
Ethoxylate (non-ionic) surfactants:
v.10: 161- (bld)
Ferrocenes: *v.5: 147*
Flavanones: 151^c (bile)
Formaldehyde: *v.10: 41 & 54 (air)*
Furfuraldehyde: *v.10: 49 (air),
235 (oils)*
Glucose: *v.10: 169 (bld)*
Glycols (as metabolites; e.g. HMPG):
52 (urn), 93 (CSF), 155^c (CSF)
Guaiphenesin: 257[1] (bld)

Hydrocarbons, various (see *also*
Benzene, Polynuc...., etc.): 154^c
(urn); *v.10: 41 (air)*, 73 & 97
(aq)

Ketones, various: *v.10: 41 (air)*
3'-O-Methylcatechin: 151^c (bile)
Naphthalene: 156^c (tss), 157^c (shrimp)
 - *also* see Carbaryl, #Ia
1-Naphthol: 7^c (urn)
Oestrogens (see *also* Steroids):
 - oestradiol: 151^c (urn), 155^c (bile),
  156^c (bld, urn), 257^c (urn), 260^c (urn)
 - synthetic: 155^c (urn, sld)

Patulin: *v.10: 114 (sld)*
Phenolphthalein: 151^c (bile, sld)
Phenol(s) (see *also* Carboxylic....
phenolic): 151^c (tss), 156^c (tss);
v.10: 25 (urn), 99 (aq)
Polynuclear aromatic hydrocarbons
(PNA, PAH; see *also* Benzo....):
151^c & 153^c & 156^c & 257[1] (tss);
v.10: 99 (aq), 115-(sld)
Quercetin: *v.5: 31 (bld, urn)*
Quinizarin: *v.10: 232-(gas oil)*

Steroid hormones (see *also* Oestrogens, & *v.7 Compound Index*): 10^c,
151^c (tss), 152^c, 154^c (bld, sld),
155^c (urn), 156^c (bile), 257^c (urn),
260 (bld), 263; *v.7: 79 (bld), 82 (urn)*
 - synthetic (see *also* #Iz, & Diethyl-
above):155^c (bile), 257[1] (bld, urn;
incl. spironolactone); *v.5: 82 (bld)*

#Iy, continued

Styrene: 153[1],^{c} (tss, urn); *v.10:
41 etc. (air)*

Vitamin D$_2$: 151^c (bile)
Vitamin K$_1$: 198[1],^{c}
Warfarin: 9[1] (bld, urn), 197, 274; *v.5:
29*

_____

#Iz [with halogen/P/N (e.g.
amide *or* imido *or* nitro);
see *also* Chloro- & Pesticides *in General Index*]

Acetaminophen (Paracetamol): 4-^{c}
(urn), 9^c (tss), 14^c, 42-(bld),
59 (bld), 156^c (tss), 157^c (tss);
v.5: 242 (bld, urn)
Acetanilide: 14[1],^{c}; *v.5: 242 (bld,
urn)*
2-Acetylaminofluorene (AAF): 257[1]
(tss), 264[1]
Acrylonitrile: 153^c;   *v. 10: 48
& 67 (air)*
Amygdalin: *v.10: 172*
Arctons: *v.7: 286*
Arochlor 1260: *v.10: 91 (aq)*

Barbiturates: *v.7: 13 (bld)*,
301
 - phenobarbital: 260 (bld);
  v.10: 183 (bld, slv , 337 (bld
 - thiopentone: 91 (bld)
Benzylisothiocyanate: 257^c (tss)

Carbon tetrachloride and related
compounds: see Carbon monoxide
& Chlor- *in General Index*
Chloral: *v.5: 138[1],^{c} (urn)*
Chlorpropamide: *v.10: 182 (bld,
slv)*
Cinnamonitrile: 153^c (urn)
Crotononitrile: 153^c (urn)

Dieldrin: 130-[1] (bld, sld, urn);
v.10: 99 (aq)
Diethylallylacetamide: 264[1] (bld)
p-Dioxins, Polychlorinated di-
benzo: *v.7: 161- (tss)*

RETRIEVAL REMINDER (Guide: p. 279): Analytes indexed according to the parent molecule's name (or group, e.g. Steroids) and structure: ten divisions, analytically meaningful - but 'aberrant' for metabolites[1], [1], [c]).

CATEGORY II (amino, not in a ring)

#IIa (acid/ester group *also*, *other than conjugate*)

RETRIEVAL REMINDER (Guide: p. 279):

Analytes indexed according to the
parent molecule's name (or group,
e.g. Steroids) and structure: ten
divisions, analytically meaningful -
but 'aberrant' for metabolites (1,$\underline{1}$,c).

SUMMARY OF CLASSIFICATION

	I	II	III
Amino (not in a ring)?	no	✓	no
Non-imide hetero-N?	no	maybe	✓
Acid or ester (not conjugate)?	✓ = Ia	✓= IIa	✓= IIIa
– no!	Halo, P or N?	Primary amino?	Hetero atom besides N?
	– no: Iy	✓ = IIb'	– no = IIIy,
	– ✓ = Iz	If no: 2^y=IIb" 3^y or 4^y =IIb'''	✓= IIIz

Only **parent compound** listed; marked [1] if Phase I metabolite(s) [$\underline{1}$ if *N*-de(s)alkylated] & ^{c} if intact conjugate(s) studied. **See also** pp. 271-2.

General Index

The entries in this Index, concerned especially with pheno-mena and points of technique, have been made comparable with those in the Indexes to Vols. 5, 7 & 10 so as to facilitate com-parison. The Index lists a few classes of analytes, but not any individual ones, nor applications of techniques such as GC.

Page entries such as 25- signify that the ensuing pages are also relevant, i.e. the - denotes a major entry.

Corrections to Vol. 10, *Trace-Organic Sample Handling*

Facing title p.: publication year for Vol. 10 **should read** 1981

p. 28, line 17: chronic **should read** chromic

p. 116, legend: stopped working **should read** working

p. 206, line 9: minimal **should read** nominal

p. 208, lines 1 & 3 of 3rd para.: after 10^{-6} & 10^{-5} **add** sec

p. 218, line 10: after ranitidine **add** to its

p. 223, line 9 from foot: pyrurate **should read** pyruvate

p. 281, legend: FCD **should read** FID

p. 323 – line 2: (s_y) **should read** (s_Y)
line 6: between is and not **insert** often
legend to Fig. 2(b): S_{yx} **should read** s_{yx}

p. 326, line 9: 90% **should read** 99%

p. 329, line 4: y in numerator **should read** Y

p. 333, lines 8, 7, 6 & 4 from foot: $s_{x'}$ **should read** $s_{\chi'}$

Compound-type Index: minor errors (e.g. the Arctons entry on p. 374 **should read** *v.7: 286*) have been eradicated in the present up-dated version (pp. 371–379, this vol.); note that in classification guides amide **should read** imide [p. 373, after amide/ (which is correct); p. 377]

General Index: to Mixing entry **add** 230– ; in Solvent..– rendered two-phase entry, **delete** 371; in Silanization and Vessels entries, 372 **should read** 371

Corrections to Vol. 11, *Cancer-Cell Organelles*

Facing title p.: publication year for Vol. 11 **should read** 1982

p. 15, Julia Roberts entry, & p. 16, Whur entry: **should read** 70–73

p. 76, line 1 of para. 2: [8] **should read** [9]

p. 115, line 15 from foot: lebulinic **should read** laevulinic

p. 119, line 6: a **should read** or

pp. 145–146: the first Plate is the one referred to in legend as *left*

p. 179, line 9 from foot: 10.25 vol. **should read** 0.25 vol.

p. 288, line 3 from foot: 30 μm **should read** 30 μM

p. 341, ref. 1: the first author **should read** El-Aaser, A.A.

p. 404, title near foot: #NC(F)-4 **should read** #NC(F)-1